PROTEIN BIOCHEMISTRY, SYNTHESIS, STRUCTURE AND CELLULAR FUNCTIONS

PROTEIN-LIPID INTERACTIONS

PERSPECTIVES, TECHNIQUES AND CHALLENGES

PROTEIN BIOCHEMISTRY, SYNTHESIS, STRUCTURE AND CELLULAR FUNCTIONS

Additional books in this series can be found on Nova's website under the Series tab.

Additional e-books in this series can be found on Nova's website under the eBooks tab.

PROTEIN BIOCHEMISTRY, SYNTHESIS, STRUCTURE AND CELLULAR FUNCTIONS

PROTEIN-LIPID INTERACTIONS

PERSPECTIVES, TECHNIQUES AND CHALLENGES

ANGEL CATALÁ
EDITOR

nova science publishers
New York

Library of Congress Cataloging-in-Publication Data

ISBN: 978-1-53613-125-3

Published by Nova Science Publishers, Inc. † New York

CONTENTS

PREFACE

Protein-Lipid Interactions: Perspectives, Techniques and Challenges analyzes and covers a broad spectrum of protein-lipid interactions and presents new information in this area of research. The topics analyzed include: an update on the cholesterol effects of the nicotinic acetylcholine receptor; the modulation of lipid compositions for protein–lipid interactions, membrane dynamics and their structures; defining the moderate affinity of specific lipid binding sites on amphitropic enzymes; the lipid-dependent gating of ion channels; the use of novel imaging approaches to study phosphoinositide signaling in vesicle trafficking; the study of protein-lipid interplay in vital biological functions; and particularly important instances of protein–lipid interactions from 1900 to 2017.

Chapter 1 - The pentameric ligand-gated ion channels (pLGIC) constitute a superfamily of neurotransmitter receptors characterized by decoding the neurotransmitter signal into a selective and rapid ion permeation process in the target cell. The lipid environment in which these receptor proteins are embedded plays an important role in modulating their topography and function. This occurs at multiple levels, including receptor trafficking, sorting, distribution and residence lifetime at the cell surface, internalization, and recycling. Cholesterol is a major player in these equilibria, via non-selective physical mechanisms operating on the bulk bilayer and through direct binding to the receptor proteins, the latter resulting in more specific influences on function. This review discusses these subjects using the crosstalk between the paradigm pLGIC, the nicotinic acetylcholine receptor, and its lipid milieu, to illustrate the influence of cholesterol on receptor structural and functional properties. The mutual interaction of the two partners is discussed at various levels of organization, from the more general influences on the lipid phase to the more intimate relationship of the sterol with the annular and non-annular sites on the receptor.

Chapter 2 - The interactions of proteins with membrane surfaces are important to many biological processes, including signaling, recognition, and membrane trafficking, cell division and cell structure. When the Fluid Mosaic Model (FMM) of membrane structure was introduced by Singer and Nicolson (1972), it was envisaged as a basic model for cell membranes that could clarify existing observations on membrane proteins and lipid structures and their dynamics. Accordingly, the membrane was topologically defined as a biological fluid of proteins and lipids oriented in two dimensions. In the FMM, amphipathic phospholipids are oriented in a lamellar mesophase organization with hydrophobic fatty acyl chains embedded within the interior of the membrane and the hydrophilic polar groups facing the aqueous environment. However, several biological processes cannot be explained on the basis of this typical phospholipid orientation but exhibit other conformations. The Lipid Whisker Model (LWM) is an extension of the FMM, introduced upon observations into the conformation of oxidized phospholipid (oxPL) species recognized by CD36. Thus, in the LWM, when cell membranes undergo oxidation, if not adapted by the action of phospholipases, they may "produce whiskers" including a variety of oxidized sn-2 fatty acids of diverse structures. In the (LWM), the assembly of many oxPL within cell membranes is different compared with one observed in non-oxPL described in the FMM. Indeed, biophysical evidence indicates that addition of an oxygen atom to the acyl chain produces significant changes that prevent its immersion in the interior of the membrane, because the presence of peroxyl radicals project toward the aqueous interface. However, controversies exist on the validity of the validity of this "floating peroxyl radical" hypothesis. This has been discussed in an opinion article.

Chapter 3 - The interaction of amphitropic enzymes with lipids is critical for transient membrane anchorage and, in many cases, for controlling enzymatic activity. While there are often crystal structures for these proteins, the likely regions interacting with membranes are often flexible loops whose orientation could easily change in the membrane-bound state. Many lipid molecules, either the same or different types, can also interact with a protein on the membrane with a wide range of lifetimes. In the case of enzymes, lipids occupying specific sites on the protein could also act as allosteric effectors of enzyme activity by promoting critical conformational changes. Thus, characterizing the membrane interactions of amphitropic enzymes has two distinct components: (1) measuring bulk partitioning of the protein onto an interface, and (2) elucidating specific interactions of lipids that affect surface binding of the protein and can also lead to conformational changes that affect catalytic activity. Bulk binding or partitioning of the protein onto the membrane can be measured with a wide variety of methods. However, identifying where and how many specific lipid binding sites are occupied on a given protein is harder. This review will briefly cover different methods to quantify protein partitioning and lifetimes on membranes, then focus on newer biophysical approaches to identifying specific phospholipid binding sites on proteins, emphasizing high resolution (or wide-range) field cycling relaxometry using site-specifically spin-labeled proteins. Several of these methods will be illustrated with a bacterial phosphatidylinositol-specific phospholipase C enzyme to illustrate how this virulence factor likely binds to target membranes in search of its substrate.

Chapter 4 - Ion channels are important because all live cells rely on them to establish a transmembrane electrostatic potential and ion gradients across their membranes. They all are integrated inside lipid bilayers and form ion conducting pores whose open states allow passive flow of certain ions down their electrochemical gradients. The dissipative flow of ions usually sends a signal at the cost of cell's energy. In general, ion channels face two aqueous phases and a lipid phase. Via chemical interactions through physical contact, ion channels can sense changes in all three phases. Ion channels are good model systems for studying lipid-channel interactions because patch-clamp recordings of single or multiple channels can be made even though manipulation of lipids immediately around ion channels is difficult to achieve. Based on the effective distance of chemical forces and the physical size of lipids, the first layer of lipid molecules, called the annular lipids, next to the transmembrane domain of a channel dominates the lipid-protein interaction. The authors propose that the annular lipids and the channel form a functional unit and the former need to rearrange in company with the allosteric conformational change in the latter. The exchange of lipids between the annular layer and the bulk lipid phase is therefore an important factor. In consideration of the chemical interactions, there are specific interaction sites for specific lipid molecules on proteins through induced-fit, which have sub-mM or higher affinity and could be stereo-specific. Low affinity binding sites are often nonspecific and may be less sensitive to the stereo-chemistry of the lipids. Lipid-dependent gating of voltage-gated ion channels refers to channel's sensitivity to the chemical differences between phospholipids and nonphospholipids. The presence or absence of phosphate groups causes a major chemical difference in charge density and H-bonding network such that the energetic differences of a channel protein between the open and closed states may become sufficiently significant in the two groups of lipids. In this chapter, the authors will focus on the prior studies that demonstrated significant lipid-dependent gating effects on ion channels, and present a systematic thermodynamic description, which may be suitable to explaining many, if not all, of the prior observations. The authors will discuss the current technical difficulties in studying the lipid-dependent gating and possible solutions in addressing them in the future.

Chapter 5 - Phosphoinositides (PIs) contribute only a small portion of membrane lipids but are essential for cell function. Dysfunction of PIs metabolism and signaling is increasingly implicated in a number of human diseases. Seven types of PIs equilibrate dynamically in living cells and have compartmentalized distribution in various cellular organelles. Such spatial organization pattern facilitates local protein interactions for organelle-specific function in time and space. As a second messenger and the precursor of DAG-PKC/Munc13 signaling pathway, $PI(4,5)P_2$ locates mainly in the inner leaflet of the plasma membrane and participates in diverse cellular functions. Here the authors focus on the PI signaling in vesicle trafficking of secretory cells and summarize the molecular mechanisms by which PIs regulate vesicle mobilization, docking, priming, fusion, clathrin-coat formation, endocytosis, and vesicle uncoating. Moreover, the authors discuss recent cutting-edge approaches that allow investigating PI signaling in a temporal-spatially defined manner during membrane trafficking. The novel technique development compensates traditional methods in biochemistry, pharmacology, and genetics, opening a new avenue to the research of lipid signaling and regulation at physiological conditions.

Chapter 6 - Proteins and lipids play substantially different but complementary roles in plasma membranes. Proteins composed of polymeric chains of chemically and physically diverse amino acid residues capable of adopting versatile and interchangeable spatial configurations combining highly ordered, anisotropically flexible and intrinsically disordered regions are ideal molecular machines sustaining nearly every process in living cells. Virtually unlimited variability of proteins allows them to adapt to any specific function in an evolutionary manner. Membrane lipids are much less versatile and have a rather narrow range of molecular weights. Their recognized primary role in cellular membranes is that of structural elements performing the barrier function that allows cells to maintain concentration gradients and electrochemical potential differences between the cell interior and exterior, as well as providing a 2D fluid environment enabling membrane protein sorting and lateral mobility, determining their folding and spatial organization. However, such a paradigm, in which the lipid bilayer is merely a passive venue for various biological functions performed by highly specialized protein molecules, appears to be an oversimplification. An essential prerequisite of life is dynamic behavior, manifested in conformational transitions of proteins and rearrangements (demixing, changes of packaging order) of the lipid matrix. Intrinsic coupling appears to exist between these processes, and the ability of the lipid subsystem to separate into coexisting liquid phases factors into this coupling in a multifaceted and dynamic manner. In a rapidly growing number of known examples, biologically significant functions appear to be mediated or driven by the dynamic processes occurring in lipid bilayers. Thus, in the process of membrane tube fission by dynamin protein, the dynamin proved to act as a mere catalyst of the physico-chemical transition resulting in a change of the membrane topology. Some cell signaling mechanisms were shown to rely heavily on the involvement of lipids in the signal transduction cascade, e.g., through specific interactions of certain lipid species with proteins, or through directed changes of physical properties that can propagate laterally to support functional coordination with other proteins and across the membrane to mediate allosteric changes in its opposite leaflet. Various protein-lipid interactions unite lipid matrix and individual membrane proteins into an integral self-consistent plasma membrane system. Selective interactions of proteins with specific lipids, the easiest kind to experimentally resolve and analyze for biological implications, are broadly recognized and characterized rather well. However, increasingly compelling and diverse lines of experimental and physiological evidence corroborated by theoretical analysis and *in silico* modeling suggest existence and biological relevance of higher-order interactions and their ability to fill the gaps in the authors' current understanding of complex and finely regulated behavior of biological membranes. Examples of the driving forces for such regulation include hydrophobic and hydrophilic interactions, mechanical deformation and curvature mediated interactions, concordant changes of protein and lipid properties induced by change of temperature or electric field distribution, and phase-separation coupled interactions. These phenomena suggest several functional and regulatory mechanisms often disregarded in the design and analysis of *in vitro* experiments and interpretation of biological and pathophysiological findings. In addition to the mechanisms supported by direct experimental evidence and detailed calculations, some more speculative options based on plausible inferences are discussed in this review, the distinction between the two being explicitly maintained. As can be seen from the examples analyzed in the review, only a few out of the plethora of possible interaction modes and physical mechanisms appear to be essential in the case of each particular protein or its biological function, justifying the hope that manageable experimental systems and theoretical models can be designed to put these hypotheses to definitive tests.

Chapter 7 - In the last five decades, the significance of protein–lipid interactions in biological systems began to be widely explored; the most important aspects in the knowledge of protein–lipid interactions from 1969 up to now are described in this chapter.

In: Protein-Lipid Interactions
Editor: Angel Catalá ISBN: 978-1-53613-125-3

Chapter 1

THE EFFECTS OF CHOLESTEROL ON THE NICOTINIC ACETYLCHOLINE RECEPTOR: AN UPDATE

Francisco J. Barrantes*
Laboratory of Molecular Neurobiology,
Biomedical Research Institute, UCA-CONICET,
Faculty of Medical Sciences, Buenos Aires, Argentina

ABSTRACT

The pentameric ligand-gated ion channels (pLGIC) constitute a superfamily of neurotransmitter receptors characterized by decoding the neurotransmitter signal into a selective and rapid ion permeation process in the target cell. The lipid environment in which these receptor proteins are embedded plays an important role in modulating their topography and function. This occurs at multiple levels, including receptor trafficking, sorting, distribution and residence lifetime at the cell surface, internalization, and recycling. Cholesterol is a major player in these equilibria, via non-selective physical mechanisms operating on the bulk bilayer and through direct binding to the receptor proteins, the latter resulting in more specific influences on function. This review discusses these subjects using the crosstalk between the paradigm pLGIC, the nicotinic acetylcholine receptor, and its lipid milieu, to illustrate the

* Corresponding Author Email: rtfjb1@gmail.com.

influence of cholesterol on receptor structural and functional properties. The mutual interaction of the two partners is discussed at various levels of organization, from the more general influences on the lipid phase to the more intimate relationship of the sterol with the annular and non-annular sites on the receptor.

Keywords: nicotinic receptor, membrane, lipid-protein interactions, membrane proteins, lipid domains, cholesterol, pentameric ligand-gated ion channels

INTRODUCTION

Neurotransmitter receptors can be subdivided into two main categories: a) the extensive superfamily of 7-transmembrane, G-protein coupled receptors (GPCR) or metabotropic receptors and b) the superfamily of pentameric ligand-gated ion channels (pLGIC), also termed ionotropic receptors. In the former case, binding of the neurotransmitter induces the coupling of the receptor with a G protein, triggering a relatively slow signaling cascade that takes the form of a metabolic process. In the case of ligand-gated ion channels, binding of the neurotransmitter induces the rapid opening of a channel that selectively allows certain ions to permeate into the postsynaptic target cell. In spite of the crucial functional (speed of the signaling process and mechanisms involved) and structural (7 versus 5 transmembrane segments) differences between the two superfamiles of receptors, both are made of transmembrane proteins with a large part of their mass embedded in the membrane lipid bilayer, and both superfamilies share common motifs to recognize lipids, and in particular, the neutral lipid cholesterol (Baier, Fantini et al., 2011, Fantini and Barrantes 2013, Fantini, Di Scala et al., 2016).

The metabotropic GPCR employ transduction mechanisms involving the cAMP signal pathway and the phosphatidylinositol signal pathway. Upon ligand binding the GPCR undergo conformational changes that enable them to perform as guanine nucleotide exchange factors (GEF). The GPCR can then activate an associated G-protein by exchanging its bound GDP for a GTP molecule. The α-subunit of the G-protein, together with the bound GTP, further proceed to dissociate from the β and γ subunits and involve downstream intracellular signaling proteins or target functional proteins directly, depending on the specific subtype of α-subunit with which they are associated.

The neurotransmitter receptors included in the pLGIC superfamily comprise the nicotinic acetylcholine receptor (nAChR), γ-aminobutyric acid type A or C ($GABA_{A}/_{C}$) receptors, glycine receptors, the subtype 3 of the serotonin (5-HT3) receptors and glutamate-gated chloride channels (GluCl) (see reviews in (Nys, Kesters et al., 2013, Taly, Hénin et al., 2014, Sauguet, Shahsavar et al., 2015)). From a functional point of view, members of the superfamily can be divided into two distinct types: the cation-selective channels such as the nAChR and 5-HT3 receptor, and the anion-selective channels such as the glycine and the $GABA_A$ and $GABA_C$ receptors and invertebrate GluCl. These functionally different classes of ion channels mediate and/or modulate excitatory or inhibitory chemical signals, thanks not only to their distinct pharmacological properties, but also to their different anatomical and cellular localization and the connectivity of the neuronal cells involved, most especially in brain.

The amino-terminal of all receptor subunits contains extracellular halves of a pair of disulfide-bonded cysteines separated by only 13 residues and is termed Cys-loop. Biochemical and X-ray crystallography studies have led to the identification of structural homologues of the eukaryotic pLGIC in Prokaryotes (Bocquet, Prado et al., 2007, Hilf and Dutzler 2008, Bocquet, Nury et al., 2009, Hilf and Dutzler 2009, Nury, Bocquet et al., 2010) reviewed in (Nys, Kesters et al., 2013, Cecchini and Changeux 2014, Olsen, Li et al., 2014, Sauguet, Shahsavar et al., 2014, Sauguet, Shahsavar et al., 2015). Although these bacterial channels do not possess a Cys-loop and their sequence identity with other members of the LGIC is rather low, their spatial (3-D) structural similarity with the eukaryotic LGIC is quite high.

pLGIC are involved in chemical signaling between neuronal cells, and between neurons and muscle cells at the neuromuscular junction. Their alterations in the so-called "channelopathies" have been implicated in various disorders of the central nervous system, ranging from Alzheimer disease (Valles, Borroni et al., 2014), schizophrenia spectrum disorders (Lin, Hsu et al., 2014), Parkinson disease (Picciotto and Zoli 2008) and other severe neuropathological conditions. Consequently, pLGIC constitute targets for a great variety of pharmacological and clinically important drugs, including general anesthetics, anxiolytics, anticonvulsants, hypnotics, behavior- and mood-modifying substances, and a variety of drugs used in the effective treatment of epilepsies or the symptomatic amelioration of Parkinson and Alzheimer diseases.

nAChR

The nAChR is the paradigm pLGIC. It is a transmembrane protein composed of five subunits organized around a central pore, the ion channel proper (Karlin 2002, Cecchini and Changeux 2015). Each nAChR subunit contains four transmembrane segments, 20–30 amino acids in length, named M1–M4. We have proposed that the interface between the protein and lipid moieties, comprising both the lipid-exposed transmembrane portions of the nAChR protein and the nAChR-vicinal lipid, behaves as an identifiable and evolutionarily conserved domain that regulates receptor function (Barrantes 2003, Barrantes 2004, Albuquerque, Pereira et al., 2009, Barrantes 2015, Barrantes and Fantini 2016).

Muscle-type nAChRs are expressed in the peripheral nervous system and neuronal-type nAChRs in both peripheral and central nervous systems as well as in other non-neural cells such as immune cells, lymphocytes, lung epithelium and others. In brain, the nAChR is present in two principal forms: the heteropentameric receptor formed by $\alpha4$ and $\beta2$ subunits and the homopentameric receptor formed exclusively by $\alpha7$ subunits (Gotti 2006, Cecchini and Changeux 2014). The nAChR rapidly became the paradigm neurotransmitter receptor, and continues to be the prototype for the pLGIC superfamily. The 30-year improvement in the electron microscopy resolution of the *Torpedo* nAChR structure, currently at about 4 Å resolution (Klymkowsky and Stroud 1979, Zingsheim, Barrantes et al., 1982, Brisson and Unwin 1985, Unwin 1993, Unwin 2005, Unwin 2013) and the crystal structure of the water-soluble homologue of the nAChR extracellular domain (the molluscan ACh-binding protein, AChBP) by X-ray diffraction techniques (Brejc, van Dijk et al., 2001, Celie, Rossum-Fikkert et al., 2004), were also important landmarks in the structural characterization of neurotransmitter receptors and structural homologues, and have had an important impact on our knowledge of pLGIC structure.

CHOLESTEROL

At variance with the great variety of phospholipid species and variants, cholesterol is chemically unique (Barrantes 2016). Cholesterol concentration alters the physical properties of the bulk membrane and also promotes the formation of distinct membrane domains with a higher degree of order than the bulk lipid membrane, termed liquid-ordered (Lo) domains (Van Meer, Voelker et al., 2008). These domains enriched in cholesterol and sphingolipids have finite lifetimes, and are structurally and chemically different from the rest of the membrane, offering a platform with unique biophysical characteristics. The most important consequence of these transient lipid supramolecular assemblies is that they spatially and temporally confine signaling processes. Cholesterol is also capable of direct interactions with neurotransmitter receptors through consensus binding sequences like the so-called cholesterol recognition/interaction amino acid consensus motif (CRAC) or its inverted signature, CARC (Fantini and Barrantes 2009, Baier, Fantini et al., 2011, Fantini and Barrantes 2013, Fantini, Di Scala et al., 2016). The CARC motif exhibits a preference for the outer membrane leaflet and its mirror motif, CRAC, for the inner one. Some membrane proteins possess the double CARC-CRAC sequences within the same transmembrane domain. Thus, neurotransmitter receptors can be influenced by cholesterol through three non-mutually exclusive mechanisms: i) changes in bulk membrane fluidity that may influence the ability of receptors to undergo the conformational changes required to exert their function; ii) association/dissociation with/from cholesterol- and sphingolipid-rich Lo membrane domains and iii) specific interactions with cholesterol that can occur within or outside cholesterol-rich domains, via cholesterol binding sites. Whereas modulation of the membrane's physical properties by cholesterol appears to be a general mechanism affecting a broad range of membrane proteins and cell functions, direct binding of cholesterol to sites on the transmembrane regions of a protein and association with membrane domains are expected to be more specific and protein-dependent. I will next describe the intimate relationship of the nAChR with lipid domains and with cholesterol, and how these relationships modulate nAChR properties.

RELATIONSHIP OF THE NACHR WITH LIPID DOMAINS

Model membrane studies showed that cholesterol–phospholipid mixtures at high cholesterol concentrations mimicked many aspects of the phase state displayed by biological membranes rich in cholesterol. These mixtures lack a defined lipid phase transition and instead are characterized by a single phase state, the liquid-ordered (Lo) phase (Ipsen, Karlström et al., 2003), with properties between the gel and the fluid lipid phases. For low cholesterol concentrations, solid-ordered (So) or liquid-disordered (Ld) phases are observed, depending on whether the system is above or below its gel-fluid transition temperature (Tm), respectively. When the binary lipid system is at intermediate cholesterol concentrations, there is phase coexistence of So and Lo (below) or Ld and Lo (above), depending on the temperature relative to Tm.

The heterogeneity of model lipid systems is also observed in biological membranes, which are far from being homogeneous. The landscape of the plasmalemma at the mesoscopic (micron-level) and microscopic/ nanoscopic levels is characterized by lateral heterogeneities of different size and composition. The chemical composition of these structural domains is also heterogeneous, and characteristically rich in saturated fatty acids, sphingomyelins and cholesterol. Their chemistry is reflected in their physicochemical status, a characteristic Lo phase, different from that of the bulk membrane bilayer lipid, which is in a liquid-disordered Ld phase (Ingólfsson, Melo et al., 2014, Krause and Regen 2014, Rao and Mayor 2014, Sevcsik and Schütz 2016). Lo domains form, transiently and reversibly, because of favorable enthalpic interactions between high melting temperature phospholipids with predominantly fully saturated or monounsaturated acyl chains, sphingomyelin and cholesterol. These nanodomains exhibit decreased conformational (trans-gauche) freedom and, consequently, viscoelastic properties different from those of the bulk lipid bilayer. Moreover, in compartmentalizing the membrane laterally, these nanodomains modify the diffusion of transmembrane proteins. The protein composition of Lo domains is also, albeit less characteristically, distinct: Lo domains are predominantly enriched in glycosyl-phosphatidylinositol (GPI)-anchored proteins, as well as in proteins involved in signal transduction and intracellular trafficking (trimeric and small GTPases, Src family kinases) (Anderson 1998). Moreover, palmitoylation serves as a signal for the anchoring of several neurotransmitter receptors in Lo domains (Resh 2013). It should be emphasized that the existence of Lo domains has been clearly demonstrated experimentally in artificial membranes, but their presence in cell membranes is still an often controversial issue. The occurrence of ultranano-domains has recently been reported (Pathak and London 2015). They were difficult to detect due to their small size (<5 nm). Lipid domains in biological membranes not only exhibit spatial heterogeneity, but also display a wide distribution in the time domain, exhibiting lifetimes from the sub-millisecond to the tens of seconds.

In the case of the nervous system, lipid domains have been reported to be involved in neuronal cell adhesion, axon guidance, synaptic transmission, growth factor-mediated signal transduction, vesicular trafficking, and membrane-associated proteolysis (Anderson 1998, Brown and London 1998, Tooze, Martens et al., 2001, Wolozin 2001, Aureli, Grassi et al., 2015). This makes Lo domains important for neural development and neuronal function and their diseases, as is the case with neurodevelopmental complex diseases like the schizophrenia spectrum disorders or the autistic spectrum disorders. The possible contribution of lipid domains to neuronal excitability stems from their reported participation in the clustering and regulation of neurotransmitter receptors on the one hand, and in the exocytic process of neurotransmitter release on the other (Martens, Navarro-Polanco et al., 2000, Bruses, Chauvet et al., 2001, Chamberlain, Burgoyne et al., 2001, Besshoh, Bawa et al., 2005, Nothdurfter, Tanasic et al., 2013, Sevcsik and Schütz 2016). Due to the essential structural role of cholesterol in the formation of liquid-ordered lipid domains, these aspects are discussed in more detail in the following sections of the review, addressing the specific roles of cholesterol on lipid platforms inhabited by nicotinic receptors.

Using simple lipid model systems, we have used fluorescence techniques to study the propensity of cholesterol and other natural and synthetic sterols and steroids to promote the formation or disruption of Lo domains (Wenz and Barrantes 2003). We found that most compounds have an ordered-phase disrupting activity, and we attributed this to the absence of carbon side chains (androstenol or C4 sterol) whereas cholesterol and 25-OH cholesterol, having such side chains, are Lo-promoting compounds (Figure 1). The side chains confer not only hydrophobicity to cholesterol and 25-OH cholesterol, but also the ability to achieve a better alignment, with their main molecular axis with a preferred orientation perpendicular to the membrane plane, and the 3-OH group at the opposite pole of the molecule at the level of the phospholipid polar head region.

In living cells, we have focused our attention on the influence of cholesterol on the formation and stabilization of receptor assemblies at the plasmalemma. Cell-surface muscle type nAChRs are organized in micron- and submicron-sized (nano)clusters, e.g., in CHO-K1/A5 cells and in resting C2C12 myotubes, respectively. It has not been straightforward to elucidate whether these receptor aggregates are associated with cholesterol-rich domains, as suggested by early experiments in which membrane patches rich in nAChR were isolated from muscle cells and their cholesterol composition subjected to biochemical analysis (Scher and Bloch 1991). Wide-field fluorescence microscopy of neuronal α7nAChR expressed in somatic spines of chicken ciliary neurons colocalized with GM1, a ganglioside reportedly associated with lipid Lo domains (Bruses, Chauvet et al., 2001). However, attempts to explore the possible association of the nAChR with cholesterol-rich Lo domains using wide-field fluorescence microscopy and cholesterol-sensitive probes did not provide conclusive results (Scher and Bloch 1991, Kamerbeek, Borroni et al., 2013). The sensitivity and the spatial resolution of the techniques did not suffice to establish the association of nAChRs with a given type of lipid domain. When reconstituted in a sphingomyelin-cholesterol-POPC (1:1:1) model system, purified nAChR from *Torpedo* does not to appear to exhibit preference for Lo domains (Bermudez, Antollini et al., 2010). However, inclusion of sphingomyelin molecular species that generate bilayer asymmetry by enriching the sphingolipid content of the outer leaflet favor the partitioning of the nAChR in Lo domains (Perillo, Penalva et al., 2016). Using coarse-grained molecular dynamics simulations Sharp and Brannigan (Sharp and Brannigan 2017) have recently simulated the partitioning of nAChRs in domain-forming lipid mixtures. These authors indicate that, contrary to expectations, nAChR partitions into the Ld phase rich in n-3 polyunsaturated fatty acids and low in cholesterol. When nAChR is partitioned into a cholesterol-poor liquid-disordered phase, binding of annular cholesterol is not observed, but cholesterol is stable in non-annular embedded sites for some cholesterol concentrations. These authors provided a structural explanation for this observation: the more flexible Ld phase can accommodate the deformation induced by the cone-shaped nAChR. In summary, experimental and in silico data point to the plasticity of the nAChR relationship with lipid domains: the receptor can inhabit/be excluded from Lo domains ("rafts") depending on the chemical composition of its lipid microenvironment.

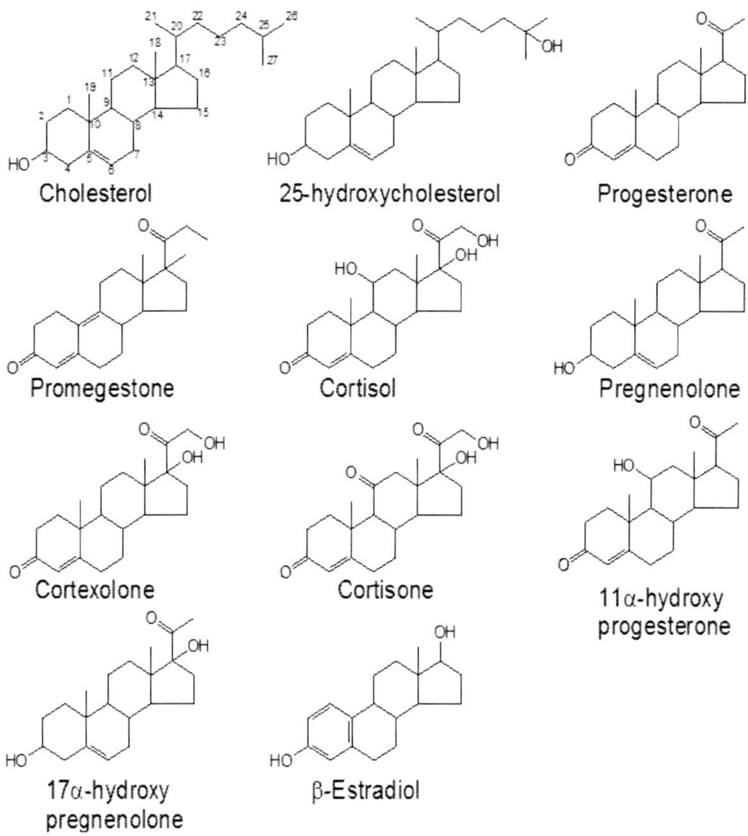

Figure 1. Sterols that promote ordered lipid domain formation/stabilization (cholesterol and 25-hydroxycholesterol) possess an aliphatic chain at C-22. This side chain might be involved in the anchoring of the steroid in the bilayer (From ref. (Wenz and Barrantes 2003).

ANNULAR AND NON-ANNULAR SITES ON THE NACHR

If we zoom in on the nAChR-lipid interactions, we begin to focus on the relationship between two types of lipids in contact with the receptor macromolecule. Annular lipids (also referred to as shell lipids or boundary lipids) are a subset of lipid molecules preferentially located at the surface of membrane-embedded proteins, in a manner analogous to the solvent layer surrounding a water-soluble protein. The surface of a membrane protein contains shallow grooves and protrusions to which the fatty acyl chains of these annular lipids adapt to provide intimate contacts. Annular lipids constitute a shell, belt or annulus which is relatively immobile, in terms of translational and rotational motions, in comparison to bulk lipids. This approximation to a definition alludes to structural as well as dynamic aspects, and implicitly to thermodynamic features of annular lipids: they exhibit affinity for binding to the transmembrane regions of the proteins, and remain in their immediate microenvironment for periods longer than those experienced by the average thermally-driven lipid molecule in the bulk bilayer. The polar head regions of annular phospholipids and the polar region of cholesterol bind to the correspondingly more hydrophilic surfaces of the membrane protein, closer to the inner and outer interfaces of the bilayer with the extracellular and intracellular moieties. The hydrophobic fatty acyl chain region of the phospholipids or the more hydrophobic part of the cholesterol molecule establishes surface contacts with the hydrophobic core of the membrane-embedded polypeptide segments.

Very few lipid molecules bound to membrane proteins have been resolved at atomic level in the scarce high-resolution structures of membrane proteins available to date. These crystallographic studies have provided evidence of the occurrence of lipid molecules between transmembrane α-helices, termed non-annular lipids. Binding of lipids to non-annular sites of some membrane proteins such as the potassium channel have been found to be functionally important (Levitan and Barrantes 2012, Rosenhouse-Dantsker, Noskov et al., 2013).

Nicotinic Acetylcholine Receptor (nAChR) and Cholesterol

The most extensively studied neurotransmitter receptor in terms of lipid-protein interactions is the nAChR (Barrantes 2003, Barrantes 2004, Barrantes 2015). Using native membrane fractions enriched in the postsynaptic membrane of *Torpedo* electrocytes interrogated with electron spin resonance (ESR) techniques, nAChR-associated lipids were shown to be immobilized with respect to the bulk membrane lipid (Marsh and Barrantes 1978). Sterols were identified as part of these immobilized lipids. Other ESR studies (Bienvenüe, Rousselet et al., 1977, Marsh, Watts et al., 1981, Marsh, Watts et al., 1982, Ellena, Blazing et al., 1983, Arias, Sankaram et al., 1990, Horvath, Arias et al., 1990, Marcheselli, Daniotti et al., 1993) corroborated the occurrence of two distinct signals in experiments with native and reconstituted membranes containing nAChR at relatively high or low concentrations: one signal corresponded to the bulk membrane lipid and the other was interpreted as stemming from the protein-immobilized lipid. These direct interactions between protein and lipid moieties were observed with fatty acids, phospholipids, and sterols in the native membrane environment. Ellena et al., (Ellena, Blazing et al., 1983) confirmed our findings using reconstituted nAChR. Rousselet and coworkers (Bienvenüe, Rousselet et al., 1977, Rousselet, Devaux et al., 1979) found immobilization with fatty acids but not with phospholipids. The Lo-type phase of the nAChR immediate perimeter was further confirmed in native membrane using fluorescence (Antollini, Soto et al., 1996) or a combination of fluorescence and single-channel patch-clamp recordings (Antollini, Soto et al., 1996). Two cholesterol pools have been described in nAChR-rich membranes from *Torpedo*: an easily extractable fraction that influences the bulk fluidity of the membrane and a tightly bound nAChR-associated fraction (Leibel, Firestone et al., 1987). These studies from different laboratories demonstrated the occurrence of different lipid pools and nAChR protein-vicinal lipids that are relatively immobile with respect to the rest of the membrane lipids. They also pointed to the existence of phase lateral heterogeneity in nAChR-rich membrane lipids earlier than the concept of “rafts” came into use in the membrane field.

Consensus motifs for cholesterol binding are present in the transmembrane domain of all nAChR subunits, rendering a total of 15 cholesterol molecules per receptor (Baier, Fantini et al., 2011). In addition, in silico modeling studies led to the conclusion that in the total absence of cholesterol the nAChR structure would collapse (Brannigan, Henin et al., 2008). This model postulates the occurrence of cholesterol sites deep inside the transmembrane region of the nAChR, a premise on which until recently there was no conclusive experimental evidence (Brannigan, Henin et al., 2008). Cholesterol sites at the lipid belt region surrounding the nAChR macromolecule, in direct contact with the membrane-facing surfaces of the transmembrane segments (Baier, Fantini et al., 2011, Fantini, Di Scala et al., 2016) have been less elusive to characterize. The free energy of interaction between cholesterol molecules and the nAChR is about -510/-530 $kJ.mol^{-1}$, i.e., more than -100 $kJ.mol^{-1}$ per subunit. The particularly favorable fit between the "CARC-like" γTM4 segment from human nAChR (428<u>R</u>VC<u>F</u>LAM<u>L</u>435) and cholesterol is noteworthy, with an energy of interaction of about -60 $kJ.mol^{-1}$, i.e., ~60% of the total energy of interaction of the entire γ subunit, which exhibits the highest affinity for cholesterol among all nAChR subunits (Figure 2). The CARC motif generally exhibits more affinity for cholesterol than the CRAC motif (Fantini and Barrantes 2013).

The significance of cholesterol for nAChR function has been extensively documented in experiments using purified nAChR from *Torpedo* reconstituted into membranes of various compositions, pointing to the clear influence this lipid –in addition to some glycerophospholipids (see reviews in (Barrantes 2003, Barrantes 2004, Barrantes 2015)) – has on the ability of the receptor to undergo the conformational transitions relevant to channel activation and desensitization. In living cells, acute cholesterol depletion has been reported to result in a *transient* ion channel gain-of-function in the cell line CHO-K1/A5 cells and in *Torpedo* nAChR expressed in *Xenopus* oocytes, whereas cholesterol enrichment has the opposite effect (Santiago, Guzman et al., 2001, Borroni, Baier et al., 2007). Subsequent studies provided a rationale for these observations: acute cholesterol depletion results in the marked acceleration of the rate of endocytosis of the receptor: the nAChR molecules remaining at the surface exhibit a moderate increase in the lifetime of their channel open state, which we interpreted as a compensatory mechanism for the temporary loss of cell-surface receptors (Borroni, Baier et al., 2007).

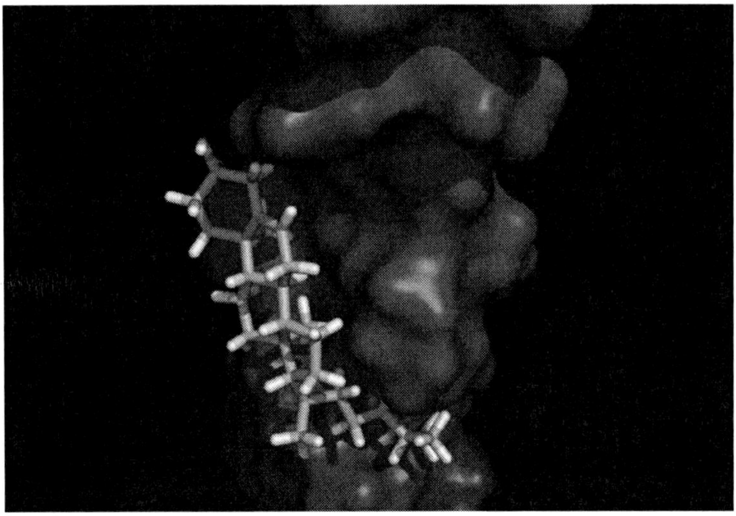

Figure 2. CARC-like domain in human AChR γTM4 transmembrane segment. This hydrophobic transmembrane segment (Swiss Prot entry P02708, surface rendering) is located in the outer membrane leaflet. γTM4 is the transmembrane segment having the highest energy of interaction (-60 kJ/mol) with cholesterol (backbone rendering).

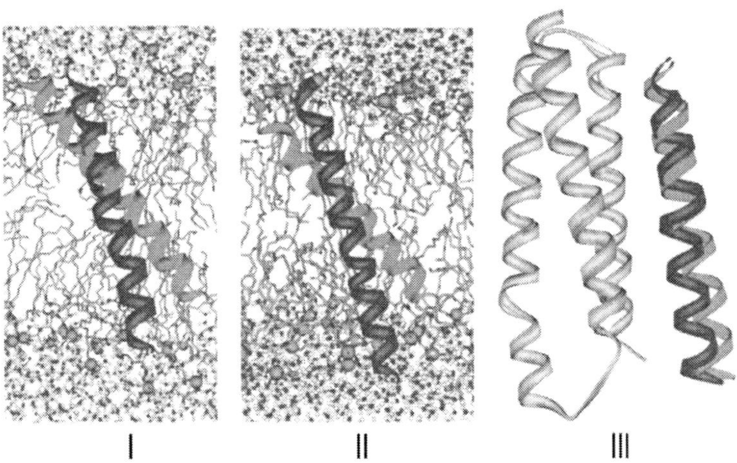

Figure 3. Changes in the orientation of I) γTM4 in POPC at 27°C, II) γTM4 in DPPC at 50°C, and III) γTM4 with the γTM1-TM2-TM3 bundle (grey) in DOPC at 50°C. From Antollini et al., (2005).

Since the function of membrane proteins is influenced by the thickness of the surrounding lipid bilayer, there is crosstalk between protein and lipid moieties to ensure matching of their respective hydrophobic surfaces. The average thickness of the bilayer-embedded portion of a membrane protein is about 29 Å. A few proteins depart from this average thickness, and generate a hydrophobic mismatch with the surrounding lipid bilayer. Membrane proteins compensate changes in membrane thickness to minimize mismatch by tilting their α-helices and rotating their side chains at the ends of the helices. In the case of the nAChR, we have established the topography of the transmembrane portions of the receptor relative to that of surrounding lipids in the bilayer by means of Förster resonance energy transfer (FRET) and differential fluorescence quenching with spin-labeled lipid analogs, sterols included (Barrantes, Antollini et al., 2000). In another series of experiments, we specifically addressed the relative topography and tilt of the representative TM4 nAChR transmembrane chain and bilayer thickness, in the presence or absence of cholesterol (de Almeida, Loura et al., 2004) or in two different phosphatidylcholines (Antollini, Xu et al., 2005). The TM4 varied its angle with respect to the membrane normal depending on the thickness of the bilayer (POPC at 27° C or DPPC at 50°C) and whether the peptide was by itself or in the whole transmembrane bundle together with TM1, TM2 and TM3 (Antollini, Xu et al., 2005) (Figure 3); but it was the presence of cholesterol that was found to affect the bilayer thickness and consequently the orientation of the reconstituted TM4 peptide more profoundly (de Almeida, Loura et al., 2004).

Cholesterol Dependence of nAChR Trafficking

There is clear experimental evidence indicating that the size, stability and organization of the nAChR at the plasma membrane depend on cholesterol levels and actin cytoskeletal integrity (Barrantes 2007, Borroni, Baier et al., 2007, Barrantes 2012, Kamerbeek, Borroni et al., 2013). Exocytic trafficking of nAChRs to the plasmalemma as well as endocytic trafficking from the cell surface to the cell interior are both cholesterol-dependent processes. In the exocytic pathway, receptors associate with cholesterol-rich domains as early as in the endoplasmic reticulum (ER) and the Golgi apparatus (Marchand, Devillers-Thiery et al., 2002). Moreover, cholesterol depletion retains nAChRs in the Golgi complex, resulting in a decrease in cell-surface nAChR levels (Pediconi, Gallegos et al., 2004). Once nAChRs reach the cell surface, their stability is also highly dependent on cholesterol levels. Acute cholesterol depletion reduces the number of receptor domains by accelerating the rate of endocytosis and shifting the internalization of the nAChR to an alternate Arf6-dependent endocytic pathway (Borroni and Barrantes 2011) different from the constitutive endocytic pathway followed by the receptor under basal conditions (Kumari, Borroni et al., 2008). In summary, cholesterol depletion leads to nAChR conformational changes that alter its stability and its long-range dynamic association with other nAChR nanoclusters, accelerates its endocytosis, and transiently affects the channel kinetics of those receptors remaining at the surface (Borroni, Baier et al., 2007).

References

Albuquerque, E. X., E. F. Pereira, M. Alkondon and S. W. Rogers (2009). "Mammalian nicotinic acetylcholine receptors: from structure to function." *Physiol Rev* 89(1): 73-120.

Anderson, R. G. (1998). "The caveolae membrane system." *Annual review of biochemistry* 67(1): 199-225.

Antollini, S. S., M. A. Soto, I. Bonini de Romanelli, C. Gutierrez-Merino, P. Sotomayor and F. J. Barrantes (1996). "Physical state of bulk and protein-associated lipid in nicotinic acetylcholine receptor-rich membrane studied by laurdan generalized polarization and fluorescence energy transfer." *Biophys J* 70(3): 1275-1284.

Antollini, S. S., M. A. Soto, I. C. Bonini de Romanelli, C. Gutierrez-Merino, P. Sotomayor and F. J. Barrantes (1996). "Physical state of bulk and protein-associated lipid in nicotinic acetylcholine receptor-rich membrane studied by laurdan generalized polarization and fluorescence energy transfer." *Biophys J.* 70(3): 1275-1284.

Antollini, S. S., Y. Xu, H. Jiang and F. J. Barrantes (2005). "Fluorescence and molecular dynamics studies of the acetylcholine receptor gammaM4 transmembrane peptide in reconstituted systems." *Mol Membr Biol* 22(6): 471-483.

Arias, H. R., M. B. Sankaram, D. Marsh and F. J. Barrantes (1990). "Effect of local anaesthetics on steroid-nicotinic acetylcholine receptor interactions in native membranes of Torpedo marmorata electric organ." *Biochim. Biophys Acta* 1027(3): 287-294.

Aureli, M., S. Grassi, S. Prioni, S. Sonnino and A. Prinetti (2015). "Lipid membrane domains in the brain." *Biochimica et Biophysica Acta (BBA)-Molecular and Cell Biology of Lipids* 1851(8): 1006-1016.

Baier, C. J., J. Fantini and F. J. Barrantes (2011). "Disclosure of cholesterol recognition motifs in transmembrane domains of the human nicotinic acetylcholine receptor." *Sci Rep* 1: 69.

Barrantes, F. J. (2003). "Modulation of nicotinic acetylcholine receptor function through the outer and middle rings of transmembrane domains." *Curr Opin Drug Discov Devel* 6(5): 620-632.

Barrantes, F. J. (2004). "Structural basis for lipid modulation of nicotinic acetylcholine receptor function." *Brain Res Brain Res Rev* 47(1-3): 71-95.

Barrantes, F. J. (2007). "Cholesterol effects on nicotinic acetylcholine receptor." *J Neurochem* 103 Suppl 1: 72-80.

Barrantes, F. J. (2012). "Regulation of the nicotinic acetylcholine receptor by cholesterol as a boundary lipid." *Cholesterol Regulation of Ion Channels and Receptors*: 181-204.

Barrantes, F. J. (2015). "Phylogenetic conservation of protein-lipid motifs in pentameric ligand-gated ion channels." *Biochim Biophys Acta* 1848(9): 1796-1805.

Barrantes, F. J. (2015). "Phylogenetic conservation of protein–lipid motifs in pentameric ligand-gated ion channels." *Biochimica et Biophysica Acta (BBA)-Biomembranes* 1848(9): 1796-1805.

Barrantes, F. J. (2016). "Cholesterol and nicotinic acetylcholine receptor: An intimate nanometer-scale spatial relationship spanning the billion year time-scale." *Biomed. Spectroscopy and Imaging_*5: S67-S86.

Barrantes, F. J., S. S. Antollini, M. P. Blanton and M. Prieto (2000). "Topography of nicotinic acetylcholine receptor membrane-embedded domains." *J Biol Chem* 275(48): 37333-37339.

Barrantes, F. J. and J. Fantini (2016). "From hopanoids to cholesterol: Molecular clocks of pentameric ligand-gated ion channels." *Prog Lipid Res* 63: 1-13.

Bermudez, V., S. S. Antollini, G. A. Fernandez Nievas, M. I. Aveldano and F. J. Barrantes (2010). "Partition profile of the nicotinic acetylcholine receptor in lipid domains upon reconstitution." *J Lipid Res* 51(9): 2629-2641.

Besshoh, S., D. Bawa, L. Teves, M. C. Wallace and J. W. Gurd (2005). "Increased phosphorylation and redistribution of NMDA receptors between synaptic lipid rafts and post-synaptic densities following transient global ischemia in the rat brain." *J Neurochem* 93(1): 186-194.

Bienvenüe, A., A. Rousselet, G. Kato and P. F. Devaux (1977). "Fluidity of the lipids next to the acetylcholine receptor protein of Torpedo membrane fragments. Use of amphopholic reversible spin-labels." *Biochemistry* 16: 841-848.

Bocquet, N., H. Nury, M. Baaden, C. Le Poupon, J. P. Changeux, M. Delarue and P. J. Corringer (2009). "X-ray structure of a pentameric ligand-gated ion channel in an apparently open conformation." *Nature* 457: 111-114.

Bocquet, N., d. C. Prado, J. Cartaud, J. Neyton, C. Le Poupon, A. Taly, T. Grutter, J. P. Changeux and P. J. Corringer (2007). "A prokaryotic proton-gated ion channel from the nicotinic acetylcholine receptor family." *Nature* 445(7123): 116-119.

Borroni, V., C. J. Baier, T. Lang, I. Bonini, M. M. White, I. Garbus and F. J. Barrantes (2007). "Cholesterol depletion activates rapid internalization of submicron-sized acetylcholine receptor domains at the cell membrane." *Mol Membr Biol* 24(1): 1-15.

Borroni, V. and F. J. Barrantes (2011). "Cholesterol modulates the rate and mechanism of acetylcholine receptor internalization." *J Biol Chem* 286(19): 17122-17132.

Brannigan, G., J. Henin, R. Law, R. Eckenhoff and M. L. Klein (2008). "Embedded cholesterol in the nicotinic acetylcholine receptor." *Proc Natl Acad Sci US* 105(38): 14418-14423.

Brejc, K., W. J. van Dijk, R. V. Klaassen, M. Schuurmans, O. J. van Der, A. B. Smit and T. K. Sixma (2001). "Crystal structure of an ACh-binding protein reveals the ligand-binding domain of nicotinic receptors." *Nature* 411(6835): 269-276.

Brisson, A. and P. N. T. Unwin (1985). "Quaternary structure of the acetylcholine receptor." *Nature* 315: 474-477.

Brown, D. and E. London (1998). "Functions of lipid rafts in biological membranes." *Annual review of cell and developmental biology* 14(1): 111-136.

Bruses, J. L., N. Chauvet and U. Rutishauser (2001). "Membrane lipid rafts are necessary for the maintenance of the (alpha)7 nicotinic acetylcholine receptor in somatic spines of ciliary neurons." *J Neurosci 21*(2): 504-512.

Cecchini, M. and J. P. Changeux (2014). "The nicotinic acetylcholine receptor and its prokaryotic homologues: Structure, conformational transitions & allosteric modulation." *Neuropharmacology*.

Cecchini, M. and J. P. Changeux (2015). "The nicotinic acetylcholine receptor and its prokaryotic homologues: Structure, conformational transitions & allosteric modulation." *Neuropharmacology* 96(Pt B): 137-149.

Celie, P. H., S. E. Rossum-Fikkert, W. J. van Dijk, K. Brejc, A. B. Smit and T. K. Sixma (2004). "Nicotine and carbamylcholine binding to nicotinic acetylcholine receptors as studied in AChBP crystal structures." *Neuron* 41(6): 907-914.

Chamberlain, L. H., R. D. Burgoyne and G. W. Gould (2001). "SNARE proteins are highly enriched in lipid rafts in PC12 cells: implications for the spatial control of exocytosis." *Proceedings of the National Academy of Sciences* 98(10): 5619-5624.

de Almeida, R. F. M., L. M. S. Loura, M. Prieto, A. Watts, A. Fedorov and F. J. Barrantes (2004). "Cholesterol modulates the organization of the gamma M4 transmembrane domain of the muscle nicotinic acetylcholine receptor." *Biophysical Journal* 86(4): 2261-2272.

Ellena, J. F., M. A. Blazing and M. G. McNamee (1983). "Lipid-protein interactions in reconstituted membranes containing acetylcholine receptor." *Biochemistry* 22: 5523-5535.

Fantini, J. and F. J. Barrantes (2009). "Sphingolipid/cholesterol regulation of neurotransmitter receptor conformation and function." *Biochim Biophys Acta* 1788(11): 2345-2361.

Fantini, J. and F. J. Barrantes (2013). "How cholesterol interacts with membrane proteins: an exploration of cholesterol-binding sites including CRAC, CARC, and tilted domains." *Front Physiol* 4: 31.

Fantini, J., C. Di Scala, C. J. Baier and F. J. Barrantes (2016). "Molecular mechanisms of protein-cholesterol interactions in plasma membranes:

Functional distinction between topological (tilted) and consensus (CARC/CRAC) domains." *Chemistry and Physics of Lipids.*

Fantini, J., C. Di Scala, L. S. Evans, P. T. F. Williamson and F. J. Barrantes (2016). "A mirror code for protein-cholesterol interactions in the two leaflets of biological membranes." *Scientific Reports* 6: 21907.

Gotti, C. R., L.; S. Vailati, S.; Clementi, F. (2006). "Brain neuronal nicotinic receptors as new targets for drug discovery." *Curr. Pharm. Des*_12: 407-428.

Hilf, R. J. and R. Dutzler (2008). "X-ray structure of a prokaryotic pentameric ligand-gated ion channel." *Nature* 452(7185): 375-379.

Hilf, R. J. and R. Dutzler (2009). "Structure of a potentially open state of a proton-activated pentameric ligand-gated ion channel." *Nature* 457(7225): 115-118.

Horvath, L. I., H. R. Arias, H. O. Hankovszky, K. Hideg, F. J. Barrantes and D. Marsh (1990). "Association of spin-labeled local anesthetics at the hydrophobic surface of acetylcholine receptor in native membranes from Torpedo marmorata." *Biochemistry* 29(37): 8707-8713.

Ingólfsson, H. I., M. N. Melo, F. J. van Eerden, C. m. Arnarez, C. A. Lopez, T. A. Wassenaar, X. Periole, A. H. de Vries, D. P. Tieleman and S. J. Marrink (2014). "Lipid organization of the plasma membrane." *Journal of the American chemical society* 136(41): 14554-14559.

Ipsen, J. H., Karlström, G, O. C. Mouritsen and M. J. Zuckermann (2003). Phase equilibria in the phosphatidylcholine-cholesterol system. *Biochim. Biophys Acta.* 905: 162-172.

Kamerbeek, C. B., V. Borroni, M. F. Pediconi, S. B. Sato, T. Kobayashi and F. J. Barrantes (2013). "Antibody-induced acetylcholine receptor clusters inhabit liquid-ordered and liquid-disordered domains." *Biophys J* 105(7): 1601-1611.

Karlin, A. (2002). "Emerging structure of the nicotinic acetylcholine receptors." *Nat Rev Neurosci* 3(2): 102-114.

Klymkowsky, M. and R. M. Stroud (1979). "Immunospecific identification and three-dimensional structure of a membrane-bound acetylcholine receptor from *Torpedo californica*." *J. Mol. Biol.* 128: 319-334.

Krause, M. R. and S. L. Regen (2014). "The structural role of cholesterol in cell membranes: from condensed bilayers to lipid rafts." *Accounts of chemical research* 47(12): 3512-3521.

Kumari, S., V. Borroni, A. Chaudhry, B. Chanda, R. Massol, S. Mayor and F. J. Barrantes (2008). "Nicotinic acetylcholine receptor is internalized via a Rac-dependent, dynamin-independent endocytic pathway." *J Cell Biol.* 181(7): 1179-1193.

Leibel, W. S., L. L. Firestone, D. C. Legler, L. M. Braswell and K. W. Miller (1987). "Two pools of cholesterol in acetylcholine receptor-rich membranes from Torpedo." *Biochim Biophys Acta* 897(2): 249-260.

Levitan, I. and F. J. Barrantes (2012). *Cholesterol Regulation of Ion Channels and Receptors*. Hoboken, NJ.

Lin, H., F. C. Hsu, B. H. Baumann, D. A. Coulter, S. A. Anderson and D. R. Lynch (2014). "Cortical parvalbumin GABAergic deficits with α7 nicotinic acetylcholine receptor deletion: implications for schizophrenia." *Mol Cell Neurosci.* 61: 163-175.

Marchand, S., A. Devillers-Thiery, S. Pons, J. P. Changeux and J. Cartaud (2002). "Rapsyn escorts the nicotinic acetylcholine receptor along the exocytic pathway via association with lipid rafts." *J Neurosci* 22(20): 8891-8901.

Marcheselli, V., J. L. Daniotti, A. C. Vidal, H. Maccioni, D. Marsh and F. J. Barrantes (1993). "Gangliosides in acetylcholine receptor-rich membranes from Torpedo marmorata and Discopyge tschudii." *Neurochem. Res.* 18(5): 599-603.

Marsh, D. and F. J. Barrantes (1978). "Immobilized lipid in acetylcholine receptor-rich membranes from Torpedo marmorata." *Proc Natl Acad Sci US* 75(9): 4329-4333.

Marsh, D., A. Watts and F. J. Barrantes (1981). "Phospholipid chain immobilization and steroid rotational immobilization in acetylcholine receptor-rich membranes from Torpedo marmorata." *Biochim. Biophys Acta* 645(1): 97-101.

Marsh, D., A. Watts, R. D. Pates, R. Uhl, P. F. Knowles and M. Esmann (1982). "ESR spin-label studies of lipid-protein interactions in membranes." *Biophys J.* 37: 265-274.

Martens, J. R., R. Navarro-Polanco, E. A. Coppock, A. Nishiyama, L. Parshley, T. D. Grobaski and M. M. Tamkun (2000). "Differential targeting of Shaker-like potassium channels to lipid rafts." *Journal of Biological Chemistry* 275(11): 7443-7446.

Nothdurfter, C., S. Tanasic, B. Di Benedetto, M. Uhr, E.-M. Wagner, K. E. Gilling, C. G. Parsons, T. Rein, F. Holsboer and R. Rupprecht (2013).

"Lipid raft integrity affects GABA A receptor, but not NMDA receptor modulation by psychopharmacological compounds." *The International Journal of Neuropsychopharmacology* 16(06): 1361-1371.

Nury, H., N. Bocquet, C. Le Poupon, B. Raynal, A. Haouz, P. J. Corringer and M. Delarue (2010). "Crystal structure of the extracellular domain of a bacterial ligand-gated ion channel." *J Mol. Biol.* 395(5): 1114-1127.

Nys, M., D. Kesters and C. Ulens (2013). "Structural insights into Cys-loop receptor function and ligand recognition." *Biochem Pharmacol* 86(8): 1042-1053.

Olsen, R. W., G. D. Li, M. Wallner, J. R. Trudell, E. J. Bertaccini, E. Lindahl, K. W. Miller, R. L. Alkana and D. L. Davies (2014). "Structural models of ligand-gated ion channels: sites of action for anesthetics and ethanol." *Alcohol Clin Exp Res.* 38: 595-603.

Pathak, P. and E. London (2015). "The Effect of Membrane Lipid Composition on the Formation of Lipid Ultrananodomains." *Biophys J* 109(8): 1630-1638.

Pediconi, M. F., C. E. Gallegos, E. B. De Los Santos and F. J. Barrantes (2004). "Metabolic cholesterol depletion hinders cell-surface trafficking of the nicotinic acetylcholine receptor." *Neuroscience* 128(2): 239-249.

Perillo, V. L., D. A. Penalva, A. J. Vitale, F. J. Barrantes and S. S. Antollini (2016). "Transbilayer asymmetry and sphingomyelin composition modulate the preferential membrane partitioning of the nicotinic acetylcholine receptor in Lo domains." *Arch Biochem Biophys* 591: 76-86.

Picciotto, M. R. and M. Zoli (2008). "Neuroprotection via nAChRs: the role of nAChRs in neurodegenerative disorders such as Alzheimer's and Parkinson's disease." *Front Biosci.* 13: 492-504.

Rao, M. and S. Mayor (2014). "Active organization of membrane constituents in living cells." *Current opinion in cell biology* 29: 126-132.

Resh, M. D. (2013). "Covalent lipid modifications of proteins." *Current Biology* 23(10): R431-R435.

Rosenhouse-Dantsker, A., S. Noskov, S. Durdagi, D. E. Logothetis and I. Levitan (2013). "Identification of novel cholesterol-binding regions in Kir2 channels." *J Biol Chem* 288(43): 31154-31164.

Rousselet, A., P. F. Devaux and K. W. Wirtz (1979). "Free fatty acids and esters can be immobilized by receptor rich membranes from *Torpedo marmorata* but not phospholipid acyl chains." *Biochem. Biophys. Res. Commun.* 90: 871-877.

Santiago, J., G. R. Guzman, L. V. Rojas, R. Marti, G. A. Asmar-Rovira, L. F. Santana, M. McNamee and J. A. Lasalde-Dominicci (2001). "Probing the effects of membrane cholesterol in the Torpedo californica acetylcholine receptor and the novel lipid-exposed mutation alpha C418W in Xenopus oocytes." *J Biol Chem* 276(49): 46523-46532.

Sauguet, L., A. Shahsavar and M. Delarue (2015). "Crystallographic studies of pharmacological sites in pentameric ligand-gated ion channels." *Biochim Biophys Acta* 1850: 511-523.

Sauguet, L., A. Shahsavar, F. Poitevin, C. Huon, A. Menny, À. Nemecz, A. Haouz, J. P. Changeux, P. J. Corringer and M. Delarue (2014). "Crystal structures of a pentameric ligand-gated ion channel provide a mechanism for activation." *Proc Natl Acad Sci US* 111: 966-971.

Scher, M. G. and R. J. Bloch (1991). "The lipid bilayer of acetylcholine receptor clusters of cultured rat myotubes is organized into morphologically distinct domains." *Exp Cell Res* 195(1): 79-91.

Sevcsik, E. and G. J. Schütz (2016). "With or without rafts? Alternative views on cell membranes." *Bio Essays* 38(2): 129-139.

Sharp, L. M. and G. Brannigan (2017). "Interactions of Nicotinic Acetylcholine Receptors with Liquid-Disordered Domains Rich in n-3 Polyunsaturated Fatty Acids." *Biophys. J.* 112: 225a.

Taly, A., J. Hénin, J. P. Changeux and M. Cecchini (2014). "Allosteric regulation of pentameric ligand-gated ion channels: an emerging mechanistic perspective." *Channels (Austin)* 8: 350-360.

Tooze, S. A., G. J. Martens and W. B. Huttner (2001). "Secretory granule biogenesis: rafting to the SNARE." *Trends in cell biology* 11(3): 116-122.

Unwin, N. (1993). "Nicotinic acetylcholine receptor at 9 Å resolution." *J. Mol. Biol.* 229: 1101-1124.

Unwin, N. (2005). "Refined structure of the nicotinic acetylcholine receptor at 4A resolution." *J Mol. Biol.* 346(4): 967-989.

Unwin, N. (2013). "Nicotinic acetylcholine receptor and the structural basis of neuromuscular transmission: insights from *Torpedo* postsynaptic membranes." *Q. Rev. Biophys.* 46: 283-322.

Valles, A. S., M. V. Borroni and F. J. Barrantes (2014). "Targeting Brain α7 Nicotinic Acetylcholine Receptors in Alzheimer's Disease: Rationale and Current Status." *CNS Drugs*.

Van Meer, G., D. R. Voelker and G. W. Feigenson (2008). "Membrane lipids: where they are and how they behave." *Nature reviews molecular cell biology* 9(2): 112-124.

Wenz, J. J. and F. J. Barrantes (2003). "Steroid structural requirements for stabilizing or disrupting lipid domains." *Biochemistry* 42(48): 14267-14276.

Wolozin, B. (2001). "A fluid connection: cholesterol and Aβ." *Proceedings of the National Academy of Sciences* 98(10): 5371-5373.

Zingsheim, H. P., F. J. Barrantes, J. Frank, W. Hanicke and D. C. Neugebauer (1982). "Direct structural localization of two toxin-recognition sites on an ACh receptor protein." *Nature* 299(5878): 81-84.

In: Protein-Lipid Interactions
Editor: Angel Catalá

ISBN: 978-1-53613-125-3

Chapter 2

PROTEIN-LIPID INTERACTIONS MODULATE LIPID COMPOSITION, MEMBRANE DYNAMICS AND STRUCTURE

*Angel Catalá**
Instituto de Investigaciones Fisicoquímicas Teóricas y Aplicadas, (INIFTA-CCT La Plata-CONICET), Facultad de Ciencias Exactas, Universidad Nacional de La Plata, Argentina

ABSTRACT

The interactions of proteins with membrane surfaces are important to many biological processes, including signaling, recognition, and membrane trafficking, cell division and cell structure.

When the Fluid Mosaic Model (FMM) of membrane structure was introduced by Singer and Nicolson (1972), it was envisaged as a basic model for cell membranes that could clarify existing observations on membrane proteins and lipid structures and their dynamics. Accordingly, the membrane was topologically defined as a biological fluid of proteins and lipids oriented in two dimensions. In the FMM, amphipathic

* Corresponding Author Email: catala@inifta.unlp.edu.ar.

phospholipids are oriented in a lamellar mesophase organization with hydrophobic fatty acyl chains embedded within the interior of the membrane and the hydrophilic polar groups facing the aqueous environment. However, several biological processes cannot be explained on the basis of this typical phospholipid orientation but exhibit other conformations. The Lipid Whisker Model (LWM) is an extension of the FMM, introduced upon observations into the conformation of oxidized phospholipid (oxPL) species recognized by CD36 (Li et al., 2007). Thus, in the LWM, when cell membranes undergo oxidation, if not adapted by the action of phospholipases, they may "produce whiskers" including a variety of oxidized sn-2 fatty acids of diverse structures. In the (LWM), the assembly of many oxPL within cell membranes is different compared with one observed in non-oxPL described in the FMM. Indeed, biophysical evidence indicates that addition of an oxygen atom to the acyl chain produces significant changes that prevent its immersion in the interior of the membrane, because the presence of peroxyl radicals project toward the aqueous interface. However, controversies exist on the validity of the validity of this "floating peroxyl radical" hypothesis. This has been discussed in an opinion article (Catala, 2015).

Introduction

The covalent connection of lipids to proteins is a basic property of all living cells. These lipidated or lipid-modified proteins are directly targeted to the membranes and exhibit various functional roles that are significant to cell function such as membrane signaling and trafficking.

1. Lipids and Lipid Binding Proteins

The importance of lipids has universally been knows during the years. Lipids are very important components of numerous biological processes and provide the building blocks of biological membranes (e.g., phospholipids, sphingolipids) or of specific proteins (e.g., myristoylation, palmitoylation), as substrate for metabolic energy production (fatty acids), and as signaling compounds (long-chain fatty acids and fatty acid metabolites). All lipid species are distinguished by their practical insolubility in aqueous solutions, i.e., their hydrophobic or amphiphilic nature, which property strictly obstructs the transport of lipids in aqueous compartments such as blood plasma, interstitium and the cellular soluble cytoplasm. But, these compartments contain proteins able of reversibly and non-covalently binding lipids - consequently designated 'lipid binding proteins' - which considerably augment the accessibility and aqueous transport of specific lipid species. These proteins include a.o. plasma albumin and cytoplasmic lipid binding proteins such as cytoplasmic fatty acid binding proteins ($FABP_c$), retinol/retinoic acid binding proteins, and oxysterol binding protein (Glatz and Van der Vusse, 1996); (Schaap et al., 2002); (Storch and Thumser, 2010). The plasma membrane in addition holds several proteins able of reversibly and non-covalently binding long-chain fatty acids, collectively named to as membrane-associated fatty acid binding proteins (Su and Abumrad, 2009); (Glatz et al., 2010); (Kazantzis and Stahl, 2012). In concert, the soluble and membrane-associated lipid binding proteins decide the bio-availability of their lipid ligands in a sequential and spatial way, and thus directly control the metabolism or signaling effect of these compounds.

The different modes by which lipid transfer proteins solubilise membrane lipids are depicted in Figure 1.

2. The Interactions of Peripheral Membrane Proteins with Biological Membranes

The composition of biological membranes is complex and varies with the type of cell or cellular compartment. Changes in lipid composition can result in the formation of micro domains with different physical properties, which arise from the interplay between the characteristic lipid headgroups and the hydrocarbon chains (Cevc, 1993). A membrane bilayer is an active environment with varying dielectric properties ranging from nonpolar within the hydrocarbon chain core to polar at the headgroup-solution interface.

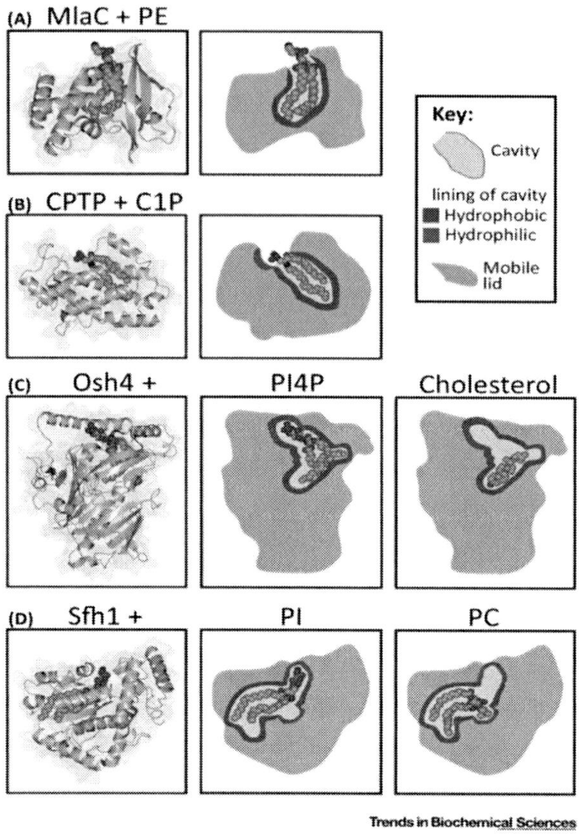

Figure 1. Different Modes by Which Lipid Transfer Proteins Solubilise Membrane Lipids. (A) MlaC (a periplasmic protein) crystallises with phosphatidylethanolamine (PE) but does not interact with its headgroup [6]. (B) Ceramide 1-phosphate (C1P) transfer protein (CPTP) binds the lipid with a hydrophilic patch outside the cavity. (C) Osh4p, a yeast relative of oxysterol binding protein (OSBP), binds phosphatidylinositol-4-phosphate (PI4P) or sterol, with two internal hydrophilic patches. (D) Sfh1, a close homologue of Sec14 in yeast, binds phosphatidylinositol (PI) or phosphatidylcholine (PC), with two internal hydrophilic patches. In both (C) and (D), the lipid is almost entirely shielded from solvent access, but these lipid transfer proteins differ in that for Osh4 (C) there are conformational changes associated with different lipid occupancy, particularly in the mobile lid. By contrast, for Sec14 and homologues including Sfh1 (D), there is no significant external response to internal occupancy. Left-hand panels: ribbon diagrams with background showing space-filling profiles and lipid ligands as space-fill format (coloured by atom: C = green, O = red, N = blue, P = magenta). Other panels: cartoons with lipid binding pockets lined according to key and major ligands: one in (A) and (B); two in (C) and (D) where the ligands shown in the ribbon diagrams are PI4P and PI, respectively. Ribbon diagrams taken from PDB files with accession numbers: 2qgu, 4k8n, 1zhy, 3spw, 3b7n, and 3b7q. Reproduced from *Trends Biochem Sci*. 2017 Jul; 42 (7): 516-530. Copyright© 2017.

In addition, the interfacial headgroup area can assume important negative charge densities, which has significant implications for recognition and recruiting membrane-associated proteins. Peripheral membrane proteins are recruited to and interrelate with cellular membranes through a sequence of different mechanisms. Specific membrane-targeting domains connect with a membrane through three major types interactions: electrostatic, hydrophobic and selective fatty acid modification. Thus, the targeted and occasionally transient interaction of a polypeptide with a bilayer is governed by a complex energy landscape.

The reason of the present review is to summarize and demonstrate mechanisms that drive the interactions between peripheral membrane proteins and membrane bilayers using precise examples from the literature. Furthermore, we emphasize common experimental approaches to probe these interactions and bioinformatics as a promising tool for predicting protein-membrane interactions. We describe how charge matching between membrane and protein surfaces give rise to long-range electrostatic interactions, how proteins become tethered to membranes by inserting hydrophobic moieties into the hydrophobic hydrocarbon regions, and how some proteins use molecular recognition to selectively target specific membrane regions or other membrane proteins.

3. The Fluid Mosaic Model of the Plasma Membrane

The typical fluid mosaic model of the plasma membrane (Singer and Nicolson, 1972) implies a homogenous allocation of lipids in the plane of the membrane bilayer. Within the structure of this model, proteins interacting with the membrane have often been viewed as islands floating in a sea of lipids. However, since the 1990s, the corresponding hypotheses of lipid domain creation and protein-mediated control of lipid packing has generated important interest in learning the lateral organization of biological membranes. Since that time, it has also become extensively recognized that proteins play significant roles in the organization of plasma membranes. Protein-lipid interactions adjust lipid composition, membrane dynamics and organization (Kozlov, 2010; Phillips et al., 2009), which are basically essential for numerous cellular processes ranging from cytoskeleton assembly (Cabeen and Jacobs-Wagner, 2005) and membrane trafficking to intracellular and intercellular signaling (Martin et al., 1998). The supramolecular assemblage rising from these protein-lipid interactions hold out a group of specific cellular functions. Thus, away from the description of individual molecular types, understanding of the highly dynamic, yet strongly controlled, interplay between membrane-associated proteins and membrane lipids is desirable to explain numerous advanced order cellular functions.

The composition of biological membranes is multifaceted and varies with the kind of cell or cellular compartment. Alterations in lipid composition can result in the creation of micro domains with different physical properties, which occur from the interaction between the characteristic lipid head groups and the hydrocarbon chains (Cevc, 1991). A membrane bilayer is an active milieu with varying dielectric properties ranging from nonpolar within the hydrocarbon chain core to polar at the head group-solution interface. Additionally, the interfacial head group area can assume important negative charge densities, which has important implications for detection and recruiting membrane-associated proteins. Peripheral membrane proteins are recruiting to and interact with cellular membranes through a series of different mechanisms. Specific membrane-targeting areas connect with a membrane through three main types interactions: electrostatic, hydrophobic and selective fatty acid alteration. Thus, the targeted and occasionally transient interaction of a polypeptide with a bilayer is governed by complex energy scenery.

Model membranes are extensively used to probe protein-membrane interactions and are thus the system of preference for biophysical studies. However, recent advances in experimental techniques, such as fluorescence microscopy, super-resolution microscopy and neutron scattering, have facilitated studies of protein-membrane interactions within the perspective of whole or live cells. Computational biology, computational resources, and a rapidly increasing number of sequenced genomes and metagenomes allow us to analyze huge amounts of sequence data by identifying relevant sequence features, such as amphipathic helices or motifs for post-translational lipid modifications. Progresses in computational chemistry have enabled exceptional insights into the interplay between peripheral membrane proteins and membranes. Although these interactions are extremely complex and dynamic, there is much to be gleaned from understanding the fundamental interaction mechanisms as they will help explain a range of cell-mediated procedures, such as signaling, membrane trafficking, and cytoskeleton dynamics, but also give insights pertinent to antibiotic functionality, drug development and probable disease mechanisms.

4. Membrane Proteins Perturb Their Environment

In a recent article (Duneau et al., 2017) describe CG molecular dynamics simulations of eleven different membrane proteins in a phospholipid bilayer and analyze them in order to explain how membrane proteins perturb their environments. The objective of this investigation was to understand how membrane proteins interact with their lipid environment, the nature of the perturbations and the relative importance of the different parameters that were proposed. Recently these authors propose a lipophobic effect (Duneau and Sturgis, 2013), in part as an extension and generalization of hydrophobic and curvature mismatch, to explain their observations of membrane protein organization (Scheuring and Sturgis, 2005). This term has also been recently used to describe various related aspects of protein-lipid interactions (Yano et al., 2015; Frotscher et al., 2015). Much of the understanding of soluble protein folding depends on our understanding of the physical-chemical basis of interactions with their solvent particularly water through the hydrophobic effect (Kauzmann, 1959) as a consequence understanding of membrane proteins will depend on our understanding of their interactions with a more complex solvent and this lipophobic effect.

5. Lipid Composition, Protein Interactions and Structure

Cellular membranes hold a great diversity of lipid species that differ in the structure of their lipid head group and in the length and saturation of their hydrophobic acyl chains. Different membrane compartments diverge both quantitatively and qualitatively in their lipid content, and there are also cell type-precise differences in lipid compositions of individual membrane compartments. For instance, the lipid composition of the plasma membrane is different between neuronal and muscle cells, and to add an additional layer of complexity, numerous signaling pathways also influence the lipid composition of the plasma membrane. In addition, whereas all lipids are moderately symmetrically dispersed between the two leaflets of the endoplasmic reticulum membrane bilayer, the Golgi, endosomal organelles, and plasma membrane exhibit an asymmetric lipid distribution, with phosphatidylcholine, sphingomyelin, and glycolsphingolipids augmented at the noncytosolic side, and phosphatidylethanolamine, as well as negatively charged phospholipids, including phosphatidylinositol and phosphatidylserine, enriched at the cytosolic leaflet (Colbeau et al., 1971; van Meer et al., 2008; van Meer and de Kroon, 2011). The asymmetric distribution of lipids has significant functional effects for numerous cellular processes, such as vesicle budding and membrane fusion (Zimmerberg and Chernomordik, 1999). In addition to lipid composition, the geometry of membranes also plays an important role in protein interactions, because many proteins preferentially interact with membranes exhibiting specific curvature (McMahon et al., 2010; Zhao et al., 2011).

Numerous factors need to be taken into account when scheming an experiment for studying membrane interactions of a specific protein. These include lipid composition, molar ratio of the lipids, membrane curvature, pH, and salt concentration of the buffer. It is significant that the lipid composition and other above-mentioned parameters in the assay will resemble as precisely as likely the relevant membrane organelle.

CONCLUSION

The interactions of proteins with membrane surfaces are important to many biological processes, including signaling, recognition, and membrane trafficking, cell division and cell structure. This review summarizes how protein-lipid interactions modulate lipid composition membrane dynamics and structure.

However, recent advances in experimental techniques, such as fluorescence microscopy, super-resolution microscopy and neutron scattering, have facilitated studies of protein-membrane interactions within the perspective of whole or live cells.

In the last five decades, the significance of protein-lipid interactions in biological systems began to be widely explored; the most important hand marks in the knowledge of protein-lipid interactions from 1969 up to now are listed below.

REFERENCES

Cabeen M. T., Jacobs-Wagner C. Bacterial cell shape. *Nat Rev Microbiol.* 2005; 8: 601-610.

Catalá Á. Lipid peroxidation modifies the assembly of biological membranes "The Lipid Whisker Model." *Front Physiol.* 2015; 5: 520.

Cevc G. Polymorphism of the bilayer membranes in the ordered phase and the molecular origin of the lipid pretransition and rippled lamellae. *Biochim Biophys Acta.* 1991; 1062: 59-69.

Colbeau A., Nachbaur J., Vignais P. M. Enzymic characterization and lipid composition of rat liver subcellular membranes. *Biochim Biophys Acta.* 1971; 249: 462-492.

Duneau J. P., Khao J., Sturgis J. N. Lipid perturbation by membrane proteins and the lipophobic effect. *Biochimica et Biophysica Acta* 2017; 1859: 126-134.

Duneau J. P., Sturgis J. N. Lateral organization of biological membranes: role of long-range interactions. *Eur Biophys J.* 2013; 42: 843-850.

Frotscher E., Danielczak B., Vargas C., Meister A., Durand G., Keller S. A fluorinated detergent for membrane-protein applications. *Angew Chem Int Ed Engl.* 2015; 54: 5069-5073.

Glatz J. F. Lipids and lipid binding proteins: a perfect match. *Prostaglandins Leukot Essent Fatty Acids*, 2015; 93: 45-49.

Glatz J. F., Luiken J. J., Bonen A. Membrane fatty acid transporters as regulators of lipid metabolism: implications for metabolic disease. *Physiol Rev.* 2010; 90: 367-417.

Kauzmann W. Some factors in the interpretation of protein denaturation. *Adv Protein Chem.* 1959; 14: 1-63.

Kazantzis M., Stahl A. Fatty acid transport proteins, implications in physiology and disease. *Biochim Biophys Acta.* 2012; 1821:852-857.

Kozlov M. M. Biophysics: Joint effort bends membrane. *Nature.* 2010; 463: 439-440.

Li X. M., Salomon R. G., Qin J., Hazen S. L. Conformation of an endogenous ligand in a membrane bilayer for the macrophage scavenger receptor CD36. *Biochemistry.* 2007; 46: 5009-5017.

Martin P. E., George C. H., Castro C., Kendall J. M., Capel J., Campbell A. K., Revilla A., Barrio L. C., Evans W. H. Assembly of chimeric connexin-aequorin proteins into functional gap junction channels. Reporting intracellular and plasma membrane calcium environments. *J Biol Chem.* 1998; 273:1719-1726.

McMahon H. T., Kozlov M. M., Martens S. Membrane curvature in synaptic vesicle fusion and beyond. *Cell.* 2010; 140: 601-605.

Phillips C. M., Goumidi L., Bertrais S., Ferguson J. F., Field M. R., Kelly E. D., Peloso G. M., Cupples L. A., Shen J., Ordovas J. M., McManus R., Hercberg S., Portugal H., Lairon D., Planells R., Roche H. M. Complement component 3 polymorphisms interact with poly-unsaturated fatty acids to modulate risk of metabolic syndrome. *Am J Clin Nutr.* 2009 90:1665-1673.

Schaap F. G., van der Vusse G. J., Glatz J. F. Evolution of the family of intracellular lipid binding proteins in vertebrates. *Mol Cell Biochem*, 2002; 239: 69-77.

Scheuring S., Sturgis J. N. Chromatic adaptation of photosynthetic membranes. *Science*. 2005; 309:484-487.

Singer S. J., Nicolson G. L. The fluid mosaic model of the structure of cell membranes. *Science*. 1972; 175: 20-31.

Storch J., Thumser A. E. Tissue-specific functions in the fatty acid-binding protein family. *J Biol Chem*. 2010; 285: 32679-32683.

Su X., Abumrad N. A. Cellular fatty acid uptake: a pathway under construction. *Trends Endocrinol Metab*. 2009; 2: 72-77.

Van Meer G., de Kroon A. I. Lipid map of the mammalian cell. *J Cell Sci*. 2011; 124: 5-8.

Van Meer G., Voelker D. R., Feigenson G. W. Membrane lipids: where they are and how they behave. *Nat Rev Mol Cell Biol*. 2008; 9: 112-124.

Yano Y., Kondo K., Kitani R., Yamamoto A., Matsuzaki K. Cholesterol-induced lipophobic interaction between transmembrane helices using ensemble and single-molecule fluorescence resonance energy transfer. *Biochemistry*. 2015; 54: 1371-1379.

Zhao H., Pykäläinen A., Lappalainen P. I-BAR domain proteins: linking actin and plasma membrane dynamics. *Curr Opin Cell Biol*. 2011; 23: 14-21.

Zimmerberg J., Chernomordik L. V. Membrane fusion. *Adv Drug Deliv Rev*. 1999; 38: 197-205.

In: Protein-Lipid Interactions
Editor: Angel Catalá
ISBN: 978-1-53613-125-3

Chapter 3

DEFINING MODERATE AFFINITY SPECIFIC LIPID BINDING SITES ON AMPHITROPIC ENZYMES

***Mary F. Roberts**, PhD**
Department of Chemistry, Boston College, Chestnut Hill, MA, US

ABSTRACT

The interaction of amphitropic enzymes with lipids is critical for transient membrane anchorage and, in many cases, for controlling enzymatic activity. While there are often crystal structures for these proteins, the likely regions interacting with membranes are often flexible loops whose orientation could easily change in the membrane-bound state. Many lipid molecules, either the same or different types, can also interact with a protein on the membrane with a wide range of lifetimes. In the case of enzymes, lipids occupying specific sites on the protein could also act as allosteric effectors of enzyme activity by promoting critical conformational changes. Thus, characterizing the membrane interactions of amphitropic enzymes has two distinct components: (1) measuring bulk partitioning of the protein onto an interface, and (2) elucidating specific interactions of lipids that affect surface binding of the protein and can also lead to conformational changes that affect catalytic activity. Bulk binding or partitioning of the protein onto the membrane can be measured

* Corresponding Author Email: mary.roberts@bc.edu.

with a wide variety of methods. However, identifying where and how many specific lipid binding sites are occupied on a given protein is harder. This review will briefly cover different methods to quantify protein partitioning and lifetimes on membranes, then focus on newer biophysical approaches to identifying specific phospholipid binding sites on proteins, emphasizing high resolution (or wide-range) field cycling relaxometry using site-specifically spin-labeled proteins. Several of these methods will be illustrated with a bacterial phosphatidylinositol-specific phospholipase C enzyme to illustrate how this virulence factor likely binds to target membranes in search of its substrate.

1. Introduction

A large number of soluble proteins can bind to membranes. These amphitropic proteins, many of which are enzymes, bind to membranes transiently, and must be carefully regulated. Signal transduction pathways have many such proteins whose membrane residence is tightly controlled, often using post-translational modifications of these soluble proteins (e.g., phosphorylation) or relying on the presence of specific lipids (phosphoinositides being the most extensively used) for spatial and temporal control of the protein binding to membranes. Amphitropic signaling proteins can have one or more very specific lipid-binding modules that could serve as a drug target in diseases [1]. Examples of these high affinity lipid binding domains (and the target lipid) include: PH (phosphoinositides [2, 3, 4] but also proteins [4, 5]), FYVE (phosphatedylinositol 3-phosphate, PI3P [2, 3, 6]), C1 (diacylglycerol [7]), and C2 (bound Ca^{2+} targets PS, but PI(4,5)P_2 can also bind to a distinct site [8]) domains. However, these specific domains need not be present. Amphitropic proteins also can have moderate affinity binding sites for more common phospholipid molecules. Identifying these moderate affinity sites is challenging but important for understanding how the protein is transiently linked to the membrane. Sometimes soluble analogues of lipid head groups or synthetic phospholipids altered to be water-soluble (e.g., phospholipids with very short acyl chains) can be used to define the specific protein-lipid interactions by crystallography or solution NMR. However, one is then defining lipid specificity in the absence of an interface, and there is no guarantee that such an interaction occurs in the more complex and constrained bilayer environment.

A different class of amphitropic proteins are those, secreted by one cell, which can initiate or alter a specific process in another cell, for example bacterial toxins. Gram positive bacteria secrete soluble enzymes that chemically modify specific phospholipids in the target cell, which in turn can cripple the response of the cell to the pathogen. The intracellular pathogen *Listeria monocytogenes* secretes a phosphatidylinositol-specific phospholipase C (*Lm*PI-PLC) that halts autophagy of the engulfed bacterium [9] allowing bacteria to proliferate inside the mammalian cell. The *Lm*PI-PLC depletes phosphatidylinositol in host membranes by converting it to diacylglycerol. If insufficient phosphatidylinositol-3-phosphate is available in the phagosomal membrane, a key step in the biogenesis of autophagosomes, lipidation of LC3, cannot occur. In a slightly different approach, *Bacillus* species, which proliferate in the intercellular milieu in mammals but also in microbial communities, secrete a structurally related PI-PLC that cleaves glycosylphosphatidylinositol (GPI)-anchored proteins from the outer leaflet of eukaryotic cell membranes [10]. Since several GPI-linked proteins participate in the innate immune response, their removal from the mammalian cell surface will prevent clearance of the pathogen by macrophages. Unicellular eukaryotes also shed GPI-anchored proteins (presumably cleaved by a PI-PLC) into their environment where they appear to modulate microbial molecular communication [11]. Directing an amphitropic enzyme to a specific intracellular or extracellular membrane will require different strategies.

The binding of amphitropic enzymes to a membrane surface is usually divided into two phases: (1) initial attraction of the protein to the target membrane, this usually has a strong electrostatic component and may only be sensitive to lipid charge rather than identity, followed by (2) two-dimensional diffusion of lipid molecules that may be substrates or modulators of enzyme activity into discrete binding sites. The first step serves to localize the protein at the surface of target membranes. Many amphitropic proteins show a general requirement for anionic lipids in a membrane for efficient protein partitioning onto interfaces, although this is not a universal requirement. If the amphitropic enzyme catalyzes hydrolysis or other chemical modification of lipids (as do the bacterial PI-PLCs), the second step involves occupation of the active site. Other discrete sites of low or moderate affinity for membrane components can also exist. These would aid in keeping the enzyme anchored on the membrane surface in the proper orientation for many rounds of catalysis. It is these weaker but still important interactions that are often the least characterized. They certainly contribute to anchoring the protein but how do you characterize them on an atomic level. This review will focus on ways to measure membrane partitioning of amphitropic proteins, to delineate the membrane-binding interface of the protein, and to characterize specific interactions of proteins with individual lipid molecules. For the last of these, particular emphasis will be placed on using high resolution shuttle field cycling to define when a specific site(s) exists.

The combination of binding studies and detailed kinetics coupled with modeling of complexes based on a variety of biophysical techniques and atomistic MD simulations can provide working models for amphitropic enzymes interacting with membranes that are testable and potentially useful for designing specific drugs that target them when cells become dysregulated.

2. Methods to Quantify Amphitropic Proteins Partitioning onto Membranes

A large number of physical techniques can measure the distribution of an amphitropic protein on a vesicle versus what remains in solution as a function of membrane concentration or specific lipid concentration. An old, but useful review [12] discusses using simple fluorescence methods, SPR, ITC, monolayer penetration and chromatographic assays to quantify amphitropic binding to membranes. Here, the emphasis will be primarily on other spectroscopic methods to characterize the membrane binding of these proteins that exist only transiently on membranes. Most of these techniques rely on a change in a measurable physical parameter that is directly (or indirectly) related to protein associating with the membrane. There are also methods to directly image binding (as long as appropriate reporter group is attached to the protein of interest).

We will illustrate several of these methods with a single relatively small amphitropic protein: *Bacillus thuringiensis* phosphatidylinositol-specific phospholipase C (*Bt*PI-PLC). This soluble enzyme cleaves GPI-anchored proteins to form a water soluble cyclic inositol moiety attached to the protein and a membrane-localized diacylglycerol molecule [13].

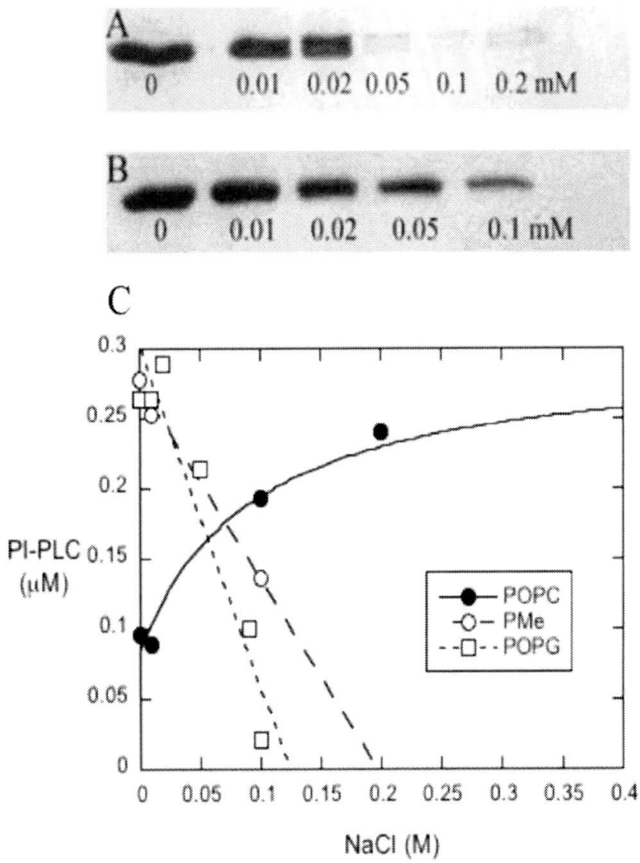

Figure 1. SDS-PAGE analysis of free *Bt*PI-PLC separated from the protein bound to SUVs of (A) 1-palmitoyl-2-oleoyl-phosphatidylglycerol (PG) and (B) 1-palmitoyl-2-oleoyl-phosphatidylcholine (PC) as a function of bulk lipid concentration (indicated beneath each lane). The total concentration of *Bt*PI-PLC used for each binding assay was 0.36 μM; buffer was 10 mM Tris HCl, pH 7.5. (C) Amount of the *Bt*PI-PLC bound to vesicles of PG (40 μM) or PMe (40 μM) as NaCl was added. Panels (A) and (B) are reprinted from Wehbi, H., Feng, J., Kolbeck, J., Ananthanarayanan, B., Cho, W., and Roberts, M.F. (2003) Investigating the interfacial binding of bacterial phosphatidylinositol-specific phospholipase. *Biochemistry* 42:9374-9382. Copyright 2003 American Chemical Society.

Phosphatidylinositol (PI) on its own is a more easily obtained substrate for PI-PLC that can be organized in bilayer vesicles as well as studied in detergent mixed micelles. The enzyme cleaves PI first to diacylglycerol and inositol 1,2-(cyclic)-phosphate (cIP). It can then hydrolyze cIP to inositol-1-phosphate [13]. Since cIP is soluble and has no tendency to partition into membranes, this substrate can be used to assess if membranes enhance the protein catalytic activity, presumably my stabilizing a particular conformation of the protein [14]. Early work with diverse compositions of phospholipids in vesicles and micelles showed that *Bt*PI-PLC catalyzed hydrolysis of cIP is activated when anionic-rich membranes contain phosphatidylcholine (PC) [15-16]. Thus, the membrane acts as an allosteric activator of *Bt*PI-PLC. How this activation occurs has been explored in depth over the last two decades.

2.1. Physical Separation

The simplest method to assess the membrane-binding ability of a protein is based solely on the size difference of the amphitropic protein in solution compared to when it is associated with lipid vesicles. There is no need for reporter labels (except if needed to enhance detection of the protein). Typically, filtration through a semipermeable membrane that retains lipid vesicles but allows free protein to flow through can be used. Centrifugation or gel filtration to separate vesicle bound protein from that free in solution is also useful. If one has no idea of binding affinity, this is an easy place to start. Initial binding studies of *Bt*PI-PLC binding to pure PC or pure phosphatidylglycerol (PG) small unilamellar vesicles (SUVs) showed that in low ionic strength media, the protein had a higher affinity for PG vesicles (Figure 1). PG or PMe was used instead of substrate PI since generation of diacylglycerol by the enzyme causes fusion of vesicles. However, as NaCl was added, the protein could be easily removed from the PG vesicles, while the presence of NaCl enhanced *Bt*PI-PLC partitioning onto the PC vesicles [17]. Another anionic phospholipid, phosphatidylmethanol (PMe), behaved much like PG. These results suggested that *Bt*PI-PLC binding to negatively charged vesicles is electrostatic and nonspecific in nature with little hydrophobic contribution, while the interaction with pure PC SUVs has a strong hydrophobic component and could be more specific.

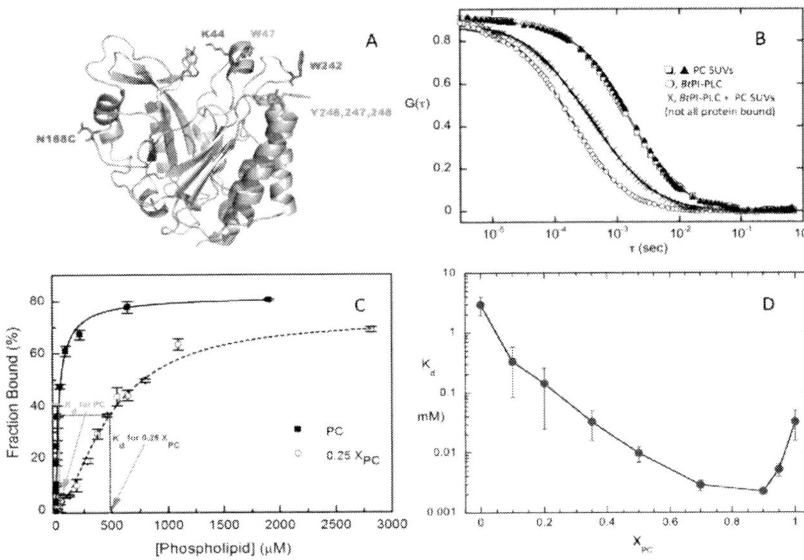

Figure 2. FCS analysis of *Bt*PI-PLC binding to SUVs as a function of mole fraction PC. (A) Structure of *Bt*PI-PLC with residues likely involved in membrane binding indicated as well as N168, the residue changed to a cysteine for attachment of the Alexa Fluor 488. (B) Cross-correlation curves for free N168C-AF488-PI-PLC (*N168C) (❍), Rho-labeled PC SUVs (❑), and Rho-labeled PMe SUVs (▲). The cross-correlation for 3.5 nM *N168C binding to 0.5 mM PC/PMe (X_{PC} = 0.50) SUVs (X), corresponding to 48% bound protein is shown. (C) Titration curves for *N168C binding to pure PC SUVs (●) and to 3:1PG/PC SUVs (❍) are shown along with the mid-points in the apparent maximum fraction bound equal to the K_d for that particular X_{PC}. (D) Dependence of *Bt*PI-PLC binding to PG/PC SUVs on X_{PC}. Both (A) and (D) are adapted from Pu, M., Roberts, M. F., and Gershenson, A. (2009) Fluorescence correlation spectroscopy of phosphatidylinositol-specific phospholipase C monitors the interplay of substrate and activator lipid binding. *Biochemistry* 48:6835-45. Copyright 2009 American Chemical Society.

These simple physical separation methods use moderately high protein concentrations (typically around 1 μM) that might complicate analyses if the protein aggregates in solution. Measuring protein binding to a wide range of vesicles as a function of lipid concentration and composition to determine apparent K_D values is tedious at best. Nonetheless, it is often a simple and reasonable place to start.

2.2. Equilibrium Fluorescence Measurements

Fluorescent methods provide a way to access lower protein concentrations than are typically used in physical separation of free and membrane-bound proteins. They can also provide dynamic information for the system. To measure the equilibrium distribution of protein onto vesicles, one uses a change in a fluorescence parameter to assess the fraction of protein bound to vesicles. Fluorescence resonance energy transfer (FRET) is the most common approach for monitoring amphitropic proteins binding to vesicles [1, 18]. FRET protein pairs have been extensively developed for following protein-protein interactions in cells [19], and the methodology is also the basis of many biosensors, including those developed for ion sensing [20], and monitoring metabolites in vivo [21].

For FRET, the donor fluorophore in an excited state (which could be tryptophan or extrinsic chromophore attached to the protein near but not compromising a membrane binding surface) must nonradiatively transfer energy to excite an acceptor fluorophore (typically either a phosphate-dylethanolamine with an appropriate fluorophore attached to the amino group or with a fluorophore on an acyl chain that is embedded in the membrane) which then fluoresces at its characteristic wavelength. For an effective donor/acceptor pair, the donor emission spectrum must overlap the acceptor absorption spectrum. This contributes to the efficiency of the FRET for a given donor-acceptor pair, as does the proximity and orientation of the two probes. Knowing the orientation of absorption and emission dipoles is critical if one wants to use the technique as an absolute molecular ruler. However, this is not very important for assessing the concentration dependence of a protein partitioning onto a membrane. One typically uses a donor-acceptor pair that is effective over long distances (10-40 Å). Typically, one measures the loss of donor fluorescence as the vesicle concentration is increased. As long as no aggregation of protein or vesicles occurs this is a quick and useful way of assessing binding. However, light scattering of larger vesicles can alter donor emission, so careful controls are needed.

Other properties of fluorophores can also be exploited. The large flat surface of many fluorophores makes them sensitive to changes in particle tumbling, which can be measured if the probe is excited with plane polarized light. The anisotropy or polarization of a probe attached to a protein will be significantly altered when it binds to the much larger lipid aggregate in the solution (e.g., bilayer vesicle, moderate size micelles, or nanodiscs). As with FRET, light scattering for larger vesicles and protein aggregation can complicate the analysis. In both experiments, while proteins are usually at much lower concentrations than the total phospholipids in vesicles, quantifying a high affinity K_d is complicated by the possibility that at very low lipid concentrations, there may not be enough vesicle surface available for the amount of protein used. The binding curve would then look sigmoidal and what one extracts is an apparent K_d. However, this can be very useful when comparing binding of a protein to different composition vesicles or in comparing a variety of different protein variants.

An alternative methodology is fluorescence correlation spectroscopy (FCS). This type of fluorescence microscopy depends on the difference in particle diffusion coefficient, D, for a protein free in solution compared to when it is bound to a phospholipid aggregate [22]. Unilamellar vesicles, which can be as small as 250 ± 50 Å diameter, can be used, and these are much larger than the size of the most amphitropic proteins free in solution [23, 24]. However, the most common vesicle size used in binding studies is around 1000 Å diameter – too large to have significant curvature on a molecular scale but not large enough to scatter light significantly and affect the fluorescence emission. A well-placed fluorophore (e.g., Alexa dyes) on the protein that is nowhere near the membrane binding site is a good reporter for these types of experiments. The very small observation volume coupled with the very low concentration of protein compared to total phospholipids (typically 1-5 nM protein with >50 nM (and usually closer to μM) total lipid), lead to at most one fluorescently labeled protein at a time on a vesicle in the very small observation volume of the microscope. These protein concentrations are typically what one might use in enzyme assays. The size distribution of the vesicles can also affect the extraction of D, particularly if the protein examined either senses curvature or defects. Evaluating D, and the fraction of protein bound, is also complicated if protein aggregation occurs or if the protein is not completely labeled. Nonetheless, we have found this to be a robust method for monitoring apparent K_d values as a function of vesicle composition and for different mutations in *Bt*PI-PLC [23, 24].

Early mutagenesis experiments in *Bt*PI-PLC indicated that Trp47 in helix B and Trp242 in a large loop at the αβ-barrel surface were critical for anchoring the enzyme to PC SUVs as measured by simple physical separation assays [25, 26]. A strip of Tyr residues in helix G also contributed to binding to PC interfaces [27]. Lys44, in the region thought to interact with the membrane, appeared important for anionic-phospholipid-rich SUVs [23]. These residues are highlighted in Figure 2A. In order to use FCS to quantify binding of the protein to PG/PC or PMe/PC vesicles as a function of the mole fraction PC (X_{PC}), a fluorophore (Alexa Fluor 488 C_5 maleimide) was covalently attached to the N168C mutation. The cysteine engineered at position 168 is far from the suspected membrane binding site, and a fluorophore at this position should easily sense changes in *Bt*PI-PLC diffusion because of the large differences in cross-correlation curves for phospholipid SUVs alone (labeled with a small amount of 1,2-dipalmitoyl-sn-glycero-3-phosphoethanolamine-N-(lissamine rhodamine B sulfonyl)) compared to free enzyme (Figure 2B). If all the protein were to bind to the vesicles, the cross-correlation curve would approach that for the SUVs alone. Treating the data for labeled protein in the presence of vesicles as a linear combination of free and SUV-bound protein generates the fraction of protein bound at a given phospholipid concentration. By varying the total phospholipid concentration, a K_d can be obtained at a given X_{PC} (Figure 2C). The K_d, expressed in terms of bulk phospholipid concentration, can then be measured as a function of X_{PC} (Figure 2D). What this showed was that membrane affinity of *Bt*PI-PLC increased about 500-fold and reached a minimum with X_{PC} ~0.80.

This 'single molecule' spectroscopy in solution method for measuring K_d values provides a direct comparison of protein/membrane binding to enzyme kinetics. Many proteins that can bind membranes tend to aggregate at the higher concentrations needed for other spectroscopies. For *Bt*PI-PLC both FCS and kinetics use similar concentration of the enzyme. Enzyme specific activity decreases for PI/PC SUVs where $X_{PC} > 0.6$. The tightest binding for the PC-rich vesicles correlates with the lowest enzyme specific activity. Such an observation could suggest that the two-dimensional K_d for substrate [28], in terms of mole fraction anionic substrate in the vesicles is, ~ 0.2. It could also be rationalized by the tight binding preventing the enzyme from dissociating into solution and finding another substrate-containing vesicle. The relative ease of using FCS to quantify binding of this amphitropic protein to SUVs of varying composition also allows one to screen for binding defects in *Bt*PI-PLC proteins with mutations around the protein face that is thought to bind to the membrane. By comparing the K_d of mutant to that of the wild type protein, we can also estimate a $\Delta\Delta G$ for the contribution of that residue to membrane binding affinity at different X_{PC} [23, 25]. This is turn can provide insight into the balance of electrostatic versus hydrophobic contributions to membrane binding [29].

A recent extension of this method that is particularly useful for protein binding to vesicles is fluorescence cross-correlation spectroscopy (FCCS) [22] where both protein and lipid populations are fluorescently labeled, and the correlation of the different wavelength emission fluorophore profiles provides a way to extract the fraction of ligand binding and the free ligand concentration that is model-independent.

2.3. Single-Molecule Fluorescence (SMF) Measurements

Equilibrium fluorescence methods reflect the balance of on and off rates for the protein binding to the vesicle. However, a mechanistic view of binding requires not only an average rate, but the range of rates in the ensemble. With solutions containing many vesicles, multiple discrete populations of binding might occur (for example if the protein binds as a monomer but can also bind as a dimer) or the protein may select out particular vesicle sizes (reflecting a possible preference of the protein for vesicle curvature). These phenomena would be averaged out with solution fluorescence methods. However, single molecule fluorescence (SMF) methods are particularly useful for assessing the dynamics of amphitropic proteins binding to bilayers [30]. For this technique, we have used phospholipid vesicles tethered to a surface and the protein of interest fluorescently labeled (again, far away from the binding surface). The fluorescence of individual proteins in solution is negligible because the molecules diffuse around so rapidly. When anchored to a bilayer attached to a surface, their fluorescence is observable. A variety of membrane interfaces can be used including supported bilayers [30, 31], tethered vesicles [32], or tethered nanodiscs [33]. As with FCS, special strategies are needed: excitation via lasers and detection with APD or CCD arrays combined with microscopy. Total internal reflection fluorescence (TIRF) is the most common set-up used. The range of residence times on membrane surfaces can be converted to off-rates, and analysis can reveal coexisting binding modes, e.g., fast and slow off-rates for a protein or peptide [31]. Unexpected insights can also be provided. Cytochrome P450 reductase is thought to exist exclusively as an integral membrane protein. However, the TIRF SMF studies on supported bilayers show that in the absence of NADPH, the protein behaves as a peripheral membrane protein [34]. Thus, substrate binding dramatically modulates the membrane binding behavior of that protein. Supported bilayers can also be used to measure D_{lat} for amphitropic proteins. In a study with proteins engineered to have one to three PH domains that bind $PI(4,5)P_2$ tightly, the lateral diffusion constant on the surface was inversely proportional to the number of PH domains, each of which presumably is bound to a phosphoinositide [30]. The tighter lipid anchors an amphitropic protein has, the tighter the binding and the slower its diffusion is in the plane of the membrane. If one wants a protein that can bind the target membrane but also be rapidly released or diffuse laterally, very tight interactions with phospholipids would not be optimal. For a good review of recent SMF uses see Hohlbein et al., [35].

For *Bt*PI-PLC, which preferentially binds to small vesicles, TIRF must be done with individual SUVs (doped with a small concentration of a fluorophore, DiD, that is used to indicate the location) tethered to a surface using biotin-avidin chemistry (Figure 3A) [32]. The fluorescence from the Alexa Fluor 488 labeled protein can then be imaged on a vesicle using a dual-view microscopy setup with a single camera that separates the two channels at 630 nm. This insures simultaneous observation of labeled *Bt*PI-PLC and DiD labeled SUVs. In the initial work, a vesicle composition rich in PC was used to ensure that the protein would bind well to the tethered vesicles. Typical time records of fluorescence from 0.2 nM *Bt*PI-PLC binding to surface tethered SUVs composed of PG/PC (3:) are show in Figure 3B. The emission for the protein fluorophore was recorded with 50 ms/frame time resolution in the example shown. For a specific SUV, each spike in the top panel of Figure 3B indicates a *Bt*PI-PLC binding to and then dissociating from that SUV. An enlargement of one particular peak, indicated by the arrow, is shown below. Each black dot indicates one frame length (50 ms). This particular peak corresponds to a residence time lasting 10 frames or 500 ms. A larger number of events were subsequently measured using tethered vesicles composed of 1:4 PG/PC. The histogram in Figure 3C shows the distribution of individual residence times for 5085 protein landings on 1168 individual SUVs. A plot of the residence times versus probability density in the inset provides the k_{off} for the interaction. This generates a lifetime of the *Bt*PI-PLC on these PC-rich SUVs equal to 379 ± 49 ms. The k_{cat} for cleavage of 1:4 PI/PC SUVs is 15-25 s^{-1}. Assuming the enzyme would have similar behavior on comparable vesicles composed of PC and PI, this timescale translates to a single enzyme potentially catalyzing the cleavage of 6-9 PI molecules on a given vesicle before dissociating.

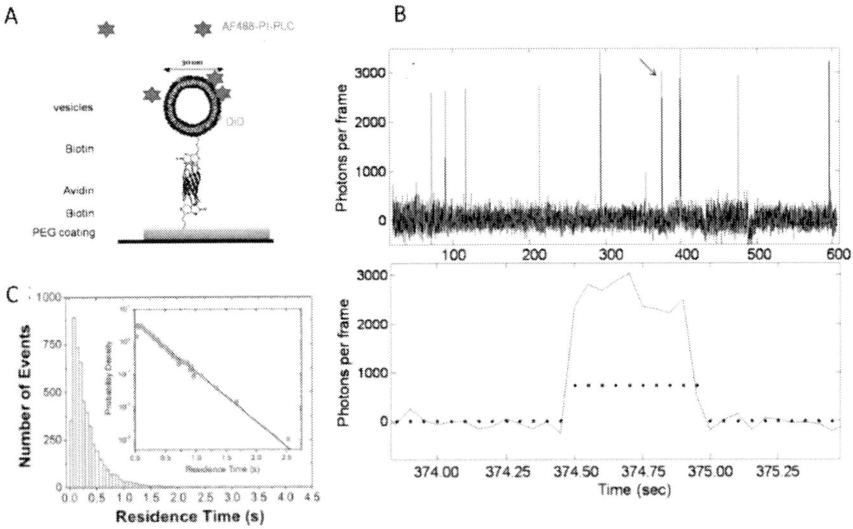

Figure 3. Single molecule fluorescence analysis of *Bt*PI-PLC binding to PC-rich tethered SUVs. (A) Schematic showing vesicle immobilization. SUVs containing biotin-PE are immobilized via neutravidin on a PEG surface containing 1% biotin-PEG. The vesicles also contain DiD in order to image the SUVs wherther or not a protein is bound. (B) Protein fluorescence intensity over time time for a single tethered vesicle (PC:PG = 7:3). Each spike corresponds to single protein landing, and the width of the spike provides the residence time, an example of which is shown in the bottom panel. (C) Analysis of 5085 protein landings on 1168 individual SUVs generates the distribution of residence times shown. These data can then be converted to k_{off} for the protein. The plot shown in (C) is reproduced with permission from the American Chemical Society from: Yang, B., Pu, M., Khan, H., Friedman, L., Reuter, N., Roberts, M. F., and Gershenson, A. (2015) Quantifying transient interactions between *Bacillus* phosphatidylinositol-specific phospholipase C and phosphatidylcholine-rich vesicles. *J. Am. Chem. Soc.* 137:14-7. Copyright 2015 American Chemical Society.

2.4. Other Spectroscopic Methods to Measure Protein Partitioning onto Vesicles

Surface second harmonic generation (SHG) spectroscopy is particularly good for monitoring the binding of macromolecules to flat surfaces. This is a coherent, nonlinear optical technique that can detect low concentrations of molecules. A recent review on the method by Tran et al., [36] provides applications aimed primarily at exploring biosensors. Attenuated total reflection Fourier transform infrared (ATR-FTIR) provides another spectroscopic approach to characterizing proteins binding to membranes. This technique also has the advantage of providing a way to orient helices that are inserted into a membrane [37] and will be discussed later.

What should be obvious in discussing all these methods is that they can assess protein binding and provide an apparent K_d for the interaction in terms of bulk phospholipid concentrations. However, what most cannot do is describe on a molecular level interactions of the protein with specific lipids.

3. Characterizing Phospholipid Binding Sites on an Amphitropic Enzyme

Amphitropic proteins anchored on a membrane are likely to have short-lifetime nonspecific interactions with many lipids in a bilayer, for example hydrophobic packing of inserted hydrophobic amino acid side chains or very transient hydrogen bonding between lipid head groups and protein residues. These short lifetime (for an individual phospholipid molecule) interactions may occlude water from a moderate size region of the protein. Added to this are specific lipid binding events that place a lipid in a well-defined site on the protein for a longer period of time. These can be referred to as specific phospholipid binding sites (not substrate binding sites if the protein is an enzyme), and they would function to hold the protein at the membrane in the correct orientation, which could involve a conformational change of the protein. There are two discrete problems to solve in describing the membrane-bound state of an amphitropic protein. (1) How is the protein positioned on the membrane (e.g., depth of insertion of any side chains in the membrane binding face), and are there significant conformational changes compared to solution or crystal structures? (2) Are there any tight and very specific phospholipid binding sites on the protein whose occupation contributes significantly to anchoring the protein at the membrane?

3.1. Orienting a Helical Protein on a Membrane

3.1.1. ATR-FTIR

Attenuated total reflection Fourier transform infrared (ATR-FTIR) spectroscopy [37] provides a way to assess the alignment of individual helices in an amphitropic protein with respect to the membrane surface. The method uses an evanescent wave at the membrane surface whose energy can only be absorbed by proteins at the membrane. Amide vibrational absorption bands are sensitive to secondary structure. If one can isolate the amide I band for a single helix in a protein transiently anchored on a membrane, measuring the dichroic ratio of ATR-FTIR spectrum provides a helix order parameter, which in turn is related to the angle of the helix relative to the membrane. For a multihelix protein, measuring the angle of two helices should provide information to align the protein on the membrane. To isolate a helix, that peptide sequence in the protein needs to be isotopically labeled (^{13}C) peptide. If the helix is near the N- or C-terminus, the protein can be expressed in the two pieces, one labeled and one not, then the two fragments connected using thioester-to-cysteine ligation. This technique should be particularly useful for amphitropic proteins where an N- or C-terminal helix is likely to interact with a membrane.

3.1.2. Solution EPR and NMR

Spin-labeling experiments, as long as the probe does not compromise membrane interactions or topology, can serve to orient the protein. When coupled with structural information for the protein alone (provided by x-ray diffraction or NMR), the EPR can be used to monitor the binding of a peptide or protein to a membrane by monitoring the peak-to-peak amplitude in the spectrum and by the deconvolution of spectra into free and bound forms [38]. The orientation/topology of the protein on membrane surface can be obtained by monitoring the effect of different relaxers on the EPR spectrum [38]. Molecular O_2 partitions into the membrane phase, while Ni(II) ethylenediamminediacetate (NiEDDA) is primarily in the aqueous phase but can also partition around the lipid polar group and the glycerol backbone. Chromium oxalate will exclusively relax spin-labels in the aqueous phase. The accessibility of spin-labels, attached covalently to a cysteine residue or incorporated as an unnatural amino acid, placed at different positons of the protein (particularly in well-defined secondary structures) provides constraints for modeling the membrane anchored structure. EPR also can be used to detect membrane-induced protein aggregates [38]. Spectra of the protein at different ratios of spin-labeled and diamagnetic protein should exhibit line narrowing as the spin-labeled protein is diluted with an unlabeled protein.

Solution NMR has the potential for generating a high resolution structure (or family of structures) in solution and then determining which residues are affected (altered chemical shift or line width) upon the addition of lipid moieties or small micelles. Amphitropic proteins cycling on and off the membrane are likely to be in fast exchange. As long as resonances have been assigned, residues that are part of a membrane binding site and/or conformational changes distinct from the membrane binding should be identifiable. Solution NMR works extremely well for small integral membrane proteins inserted into micelles or bicelles. However, it is more difficult to get a glimpse of what the membrane structure is for an amphitropic protein. In these experiments, the proteins, typically at 0.1 – 0.5 mM, often aggregate to larger structures when amphiphiles are added. This can complicate spectral changes or make them difficult to distinguish.

3.1.3. HDX-MS

A particularly useful way to identify segments of a protein that bind to membranes is hydrogen-deuterium exchange mass spectrometry (HDX-MS) [39]. The basis of the methodology is that when a protein is placed in D_2O, an amide –NH– can exchange its hydrogen atom for a deuterium from the solvent if the amide group is exposed to solvent. If the residue exists in a stable secondary structure unit (α-helix or β-sheet), the exchange occurs more slowly. If the protein segment is intrinsically disordered, the exchange is fast. Analysis of deuterium incorporation as a function of time requires enzymatic digestion of the protein samples under acidic conditions (to prevent further exchange), usually with pepsin. MS-MS spectra are used to identify peptides. What one obtains for each peptide detected is then a fraction deuteration, f_D, that can be monitored as a function of incubation in the D_2O solution.

When an amphitropic protein binds to a membrane, solvent will be partially occluded from the amides of the residues comprising the binding site, so that there is a significant change in the HDX-rate of peptides containing those residues. The difference in rate is most pronounced if a peptide is in a highly flexible or disordered region of the protein. The method can accommodate complexity – membranes with multiple lipids and different soluble molecules or ions that bind to the protein can be used. The experiment can also include proteins thought to interact with one another, either in solution or when transiently docked on a membrane. Not only can the method supply a medium resolution picture of the membrane interacting regions of an amphitropic protein, but it can identify the peptides that form such protein-protein interfaces, both in solution and in the membrane-bound complex. Defining the topology of the protein-protein binding surfaces is an important problem in understanding details of multiprotein complexes formed in signal transduction pathways or in other complex biological processes [40]. However, optimal use of the methodology requires a structure of the protein.

The αβ-barrel of *Bt*PI-PLC has two features that mutagenesis coupled with measurements of membrane binding suggest are important: (1) two surface Trp residues, one in short helix B (-IKQV^{47}WG-) and the other in a large rim loop (-SGGTA^{242}W-), and (2) a tyrosine strip at the N-terminus of helix G (^{246}YYYASY). The short helix and large loop have well-defined positions in the crystal structure of the native protein [41]. However, they may have more flexibility or multiple orientations or structures when the protein is in solution. If they are indeed in the membrane binding region of the protein their susceptibility to HDX should be reduced when the protein binds to membranes. The simplest HDX-MS experiment is to measure the H/D exchange in the absence and presence of small vesicles to which the protein binds fairly tightly. For *Bt*PI-PLC this means looking at the deuteration rate in the absence and presence of PC-rich SUVs.

In the absence of vesicles, fast amide exchange is seen for the N-terminal peptide (not shown) and the large loop between β-strand VII and α-helix G (peptide residues 232-251). β-strand V, covered by the peptide with residues 182-195, also exhibits more exchange than many other strands in the structure. Helix B, which contains Trp47 whose loss has a large effect on membrane binding, is a very stable structure with little deuterium incorporation into the peptide with residues 38-55.

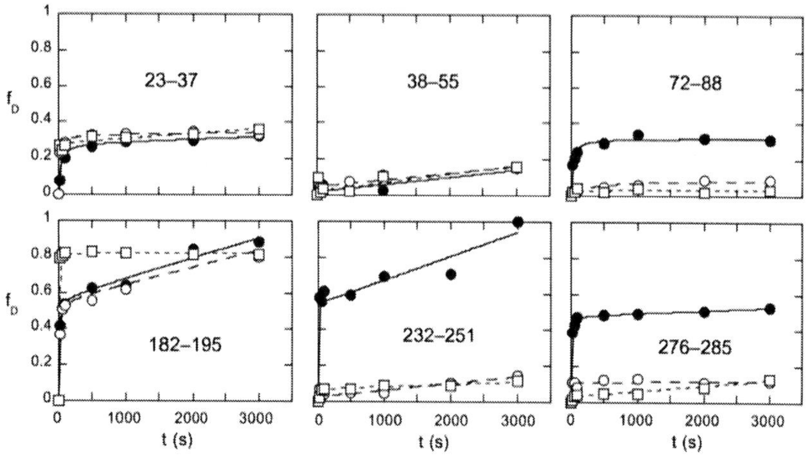

Figure 4. The fraction of deuterium incorporation, f_D, into selected peptides of *Bt*PI-PLC in the absence and presence of PC/PG SUVs. Filled circles represent the observed f_D for the protein when diluted with D_2O buffer. Changes in f_D in *Bt*PI-PLC binding to PC/PG (9:1) are indicated by squares and those incubated with PC/PG (8:2) by open circles. The data are from the PhD thesis of Dr. Xiaomeng Shi, Boston College, *Phosphatidylinositol-specific phospholipase C: conformational changes upon membrane binding*, 2010.

In the presence of the PC/PG vesicles, while many peptides show little change in exchange behavior, the peptide around helix G and the loop leading up to it (residues 232-251) are dramatically protected. We already know that Trp242, and one or more of the Tyr in helix G are likely to interact with the membrane, and this is confirmed by the HDX-MS. Helix B from other work appears membrane embedded, however, that information is not provided from the HDX-MS because in solution this short helix is quite stable.

The f_D profiles for several peptides are shown in Figure 4. Two peptides (23-37 and 38-55) show no change in the f_D pattern. Peptide 23-37 contains the first β-strand and 38-55 contains helix B. Peptide 182-195, containing β-strand V, exhibits a high initial f_D with no protective effect from vesicles. Other parts of this peptide must not adopt a stable secondary structure in the structure and the region is not part of the membrane interface (or alternatively primed by membrane binding to adopt an exchange resistant motif when the protein is bound). Peptide 72-88 shows a significant decrease in f_D in the fast phase; this peptide contains His 82, an active site residue, as well as two Tyr86 and Tyr88 that could insert into membranes. Peptide 232-251 has the large loop containing Trp242 and the initial part of helix G with the Tyr strip. In solution in the absence of SUVs about half the peptide exchanges rapidly (presumably the loop) while the rest has a slower exchange rate (helix G). This peptide shows the largest protection from deuteration when the protein binds to vesicles, a result that strongly implies that the large loop interacts with the membrane. Peptide 278-285, at the C-terminus of the protein, is a helix in the crystal structure. That about half the amides exchange rapidly may indicate that the C-terminal portion is dynamic and can partially unravel in solution. However, the exchange is reduced when the protein binds to membranes, and this segment is not very close to regions that interact with the bilayer.

This example for a small amphitropic protein shows that while the structural details of membrane binding are not provided, information is provided not only on peptides that likely interact with the membrane but also dynamic information on other parts of the protein that are altered in response to membrane binding. It also points out that stable secondary structures can have side chains that interact with membranes, but their amides will not necessarily show any change in deuteration rates. What is really needed to get the most out of this technique is a way to translate many of these other changes, particularly where exchange rates increase, into specific dynamic motions of the protein.

3.2. Identifying Specific Lipid Binding Sites on Amphitropic Proteins

Crystal structures of some integral membrane proteins exhibit well-defined sites where phospholipid head groups interact with the protein via ion pairs and hydrogen bonds. Specific lipids play functional roles in many transporters, and specific complexes have been identified [42]. Similar interactions could occur with peripheral membrane proteins. With the amphitropic proteins, soaking crystals with high concentrations of soluble segments of phospholipids (phosphomonoesters such as phosphocholine or phosphoserine or glycerophosphodiesters) can identify regions on the protein that could bind phospholipid headgroups [43, 44]. Affinities of the soluble segments alone are usually not very high and at the high concentrations added either in the crystallization media or in crystal soaks, adventitious binding could occur. Mutagenesis in the region where one suspects a ligand to bind and loss of membrane binding affinity are required. However, substitution of bulky or charged residues with alanine (often the default procedure) can alter electrostatic contributions to binding, and/or alter protein dynamics or conformation in an area distinct from any lipid binding site. These structures with pieces of phospholipids can often inform molecular dynamics studies which can monitor specific phospholipid binding by noting how long the interaction persists (although for atomistic simulations, the longest simulations only cover about 500 ns [44, 45]).

Recently, an underused NMR technique, high resolution ^{31}P field cycling relaxometry, has been shown to have the capacity to identify discrete phospholipid binding sites on proteins. The technique involves polarizing the spins in the sample in the probe of the NMR spectrometer at the high static field (B_0), then rapidly moving the sample up the bore of the magnet to a position where it will relax at a lower field (B_{relax}) for a prescribed amount of time. As B_{relax} is decreased, dipolar relaxation, a complex sum of terms related to $1/(1+\omega^2\tau^2)$, will increase and reach a maximum when $\omega^2\tau^2 << 1$. Here, ω is the radial resonance frequency of the nucleus observed (here ^{31}P) and τ is the correlation time for that particular relaxation mechanism. The sample is then rapidly moved back to the probe to read-out the residual polarization. The technique is useful because (i) polarization of the spins, which will determine experimental sensitivity, increases with magnetic field, (ii) read-out at high field will preserve the chemical shift differences for ^{31}P in different phospholipids, allowing several different phospholipids to be examined in the same vesicle, and (iii) the magnitude of the dipolar relaxation is proportional to the distance between ^{31}P and the dipole relaxing it to the sixth power. The simplest field cycling experiment uses multicomponent small unilamellar vesicles (SUVs) or phospholipids presented in micelles and focuses on the ^{31}P nuclei of the phospholipids, since these are well separated in higher field magnets. For those interested, details of the instrument [46] and initial studies of phospholipids in SUVs [47-50] and derivation of key parameters are described in detail.

Biological phosphodiester and monophosphates have moderately long ^{31}P spin-lattice relaxation times (T_1) at high fields and very little dipolar relaxation from other nearby nuclear spins (the nearest protons are three bonds away from the phosphorus atom). For these phosphates, chemical shift anisotropy (CSA) is the major relaxation mechanism at high fields, and it entails fast motions around the ^{31}P with a correlation time ~ 5×10^{-10} s (M.F. Roberts and A.G. Redfield, unpublished results). Dipolar spin-lattice relaxation becomes a significant mechanism for ^{31}P relaxation in macromolecules, phospholipid bilayers, micellar aggregates, and even small molecules only at magnetic fields below 2 T – well below the static field of conventional high field magnets (a spectrometer where the ^{1}H frequency is 500 MHz has a magnetic field of 11.7 T). For phospholipids, dipolar relaxation is useful because it reflects proximity of nearby protons to the ^{31}P of interest as well as time-scales of phospholipid motions that affect the ^{31}P-^{1}H vectors. However, it is complicated for phospholipids (except for phosphatidic acid) because the phosphodiesters have two different –CH_2– or –CH– groups three bonds away and the contributions of each to dipolar relaxation could be different. Furthermore, if a protein binds transiently to a phospholipid surface, it can potentially also contribute to the dipolar relaxation.

For phospholipids in an SUV, the dependence of spin-lattice relaxation rate (R_1 = $1/T_1$) on field shows a complex but interpretable pattern as indicated in Figure 5. A CSA contribution, R_{CSA}, is detected at high fields. R_{CSA} is proportional to ω^2, and the correlation time, τ_{CSA}, is short. Between 0.04 and 2 T a dipolar contribution with a correlation time ~10 ns is detected. For that motion one can extrapolate the maximum relaxation rate to a magnetic field of zero. As $\omega^2\tau^2$ increases above 0.1 T, the R_1 will decrease. The time-scale of that dipolar relaxation could reflect wobble-in-a-cone of the phospholipid head group [49] or effective rotation of the phospholipid in the bilayer [47]. There is also another dipolar contribution at much lower fields. For SUVs it starts to contribute to ^{31}P relaxation below 0.04 T and is characterized by a correlation time roughly on the order of the overall tumbling time of the vesicle [48].

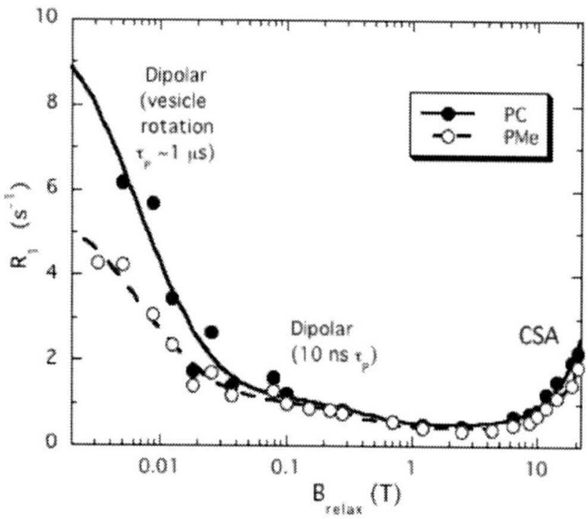

Figure 5. ^{31}P Relaxation rate (R_1) as a function of B_{relax} for PC and PMe (1:1) in small unilamellar vesicles. The three easily discernible relaxation mechanisms include the low field dipolar profile (larger for PC than PMe) that is caused by overall vesicle rotation ($\tau_P \sim 1$ μs), the mid-field dipolar contribution, caused by a local P · · · · H motion with a correlation time ~ 10 ns (similar for both phospholipids), and a high field CSA contribution, again similar for both. A small cross-correlation contribution between the low field dipolar and high field CSA is also present, but in the regions shown this should be a constant (~0.4 s^{-1}).

If one attaches a spin-label such as 1-oxyl-2,2,5,5-tetra-methylpyrroline-3-methyl-methanethiosulfonate to the protein in a well-defined place (easy for this probe because it reacts with the cysteine thiol group), then the much larger electron dipole of the spin-label dominates the ^{31}P relaxation if the two are close enough. One still wants to acquire relaxation data at low fields, and not just a single relaxation rate, in order to examine how the relaxation rate depends on B_{relax} or radial frequency ($\omega_P = \gamma_P B_{relax}$ where γ_P is the ^{31} gyromagnetic ratio).

The spin-label on a protein will enhance the relaxation rate of a ^{31}P in a phospholipid only when that protein is bound to the vesicle. The increased relaxation rate caused by the spin-label is termed the PR_1E, paramagnetic R_1 enhancement. How close the unpaired electron site is to the phospholipid will determine the strength of this ^{31}P-electron interaction. Time-scale and motions also are critical here. For the 10 ns motion, any transient complex that is stable for that short correlation time will give rise to a large PR_1E. Many phospholipids may sample the spin-label on this time scale, so rather than a well-defined ^{31}P-e distance, this dipolar dispersion senses the average distance of the unpaired electron to the ^{31}P at the membrane surface. However, for the very low field dispersion to show a large PR_1E, the complex of the protein with a phospholipid bound in a discrete site must exist for a least a μs. However, an individual phospholipid can't be bound extremely tightly for the correlation time of a 0.1-0.5 s, as suggested by the SMF lifetime of amphitropic proteins on a vesicle. That is sufficiently long to prevent fast exchange of the free and bound phospholipids, so no PR_1E would be detected.

*Bt*PI-PLC was the first amphitropic protein examined with this methodology since it was shown to be kinetically activated when PC was added to a substrate PI vesicle [15, 16]. The presence of the PC not only affected k_{cat} but FCS studies showed that it enhanced membrane binding of the enzyme by several orders of magnitude. Of course changes in the membrane properties could also alter both vesicle binding and the subsequent enzyme activity. Older experiments did suggest one or more specific sites were involved. When the protein was incubated with diC_7PC micelles (no anionic phospholipids that would bind in the active site) and then crosslinked with a bifunctional crosslinking agent, two diC_7PC molecules were found tightly associated with the protein (analyzed by MALDI-TOF MS) [51]. If there are discrete PC-binding sites on this amphitropic protein, where are they?

Figure 6 shows R_1 profiles for PC and PMe, both in the same SUV at 1:1, at very low relax B_{relax} with three different spin-labeled *Bt*PI-PLCs added at a ratio of 300 of each phospholipid to one enzyme. While an individual vesicle will likely have more than one enzyme bound transiently, it will not be covered with the protein. The PR_1E requires close proximity of lipids and proteins so that an easily measurable effect can be observed and the experiment tuned to measure a reasonable change in R_1. For each phospholipid, the profile with the spin label on N168C (the same position as used in FCS experiments) serves as a control as this position is far from the binding binding face of the protein (Figure 6 A and B). The PR_1E at each field, ΔR_1, is computed by subtracting the R_1 with unlabeled protein from the R_1 observed with spin-labeled protein. Examining the profiles for each phospholipid in Figure 6, we immediately see that the spin-label on *Bt*PI-PLC that is closest to the PMe, which binds at the active site, is attached to W47C in short helix B. The spin-label at D205C is further away, but a small PR_1E is observed. The spin-label on D205C is the closest to a discrete PC molecule, while spin-labeled W47C is virtually indistinguishable from the control (spin label on N168C).

The ΔR_1 field dependence for each phospholipid in the presence of a spin-labeled protein is shown in Figure 6 C and D. The ΔR_1 field profile is then fit with Equation 1,

$$\Delta R_1 = \Delta R_{P\text{-}e}(0)/(1 + \omega_P^2 \tau_{P\text{-}e}^2) + c \quad (1)$$

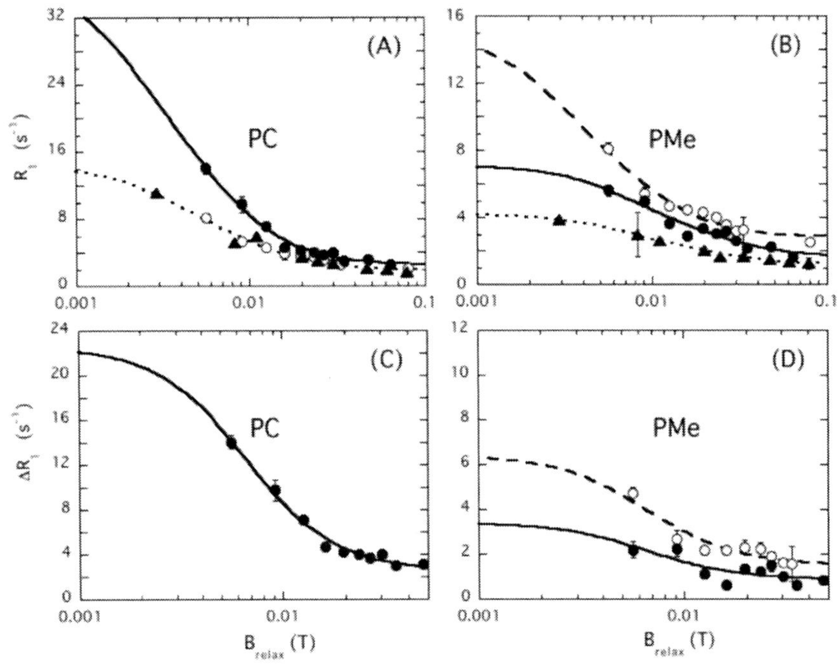

Figure 6. Low field dependence of the ^{31}P R_1 for PMe/PC vesicles in the presence of spin-labeled *Bt*PI-PLC. (A) PC or (B) PMe in mixed vesicles of PC/PMe (5 mM: 5 mM) in the presence of 14.4 μM *Bt*PI-PLC spin-labeled on a single cysteine: W47C (❍), D205C (●), and N168C (▲). (C) Field dependence of the PR_1E (ΔR_1) for PC due to spin-labeled D205C protein. The optimized fit is consistent with the $\tau_{P\text{-}e} = 1.4 \pm 0.2$ μs; $\Delta R_{P\text{-}e}(0)$, the other key parameter for determining the P-e distance, is 20 ± 2 s^{-1}. (D) Field dependence of ΔR_1 for PMe in the presence of the same spin-labeled *Bt*PI-PLC enzymes. For these anionic phospholipids and the same $\tau_{P\text{-}e}$, $\Delta R_{P\text{-}e}(0)$ is 4.9 ± 0.7 s^{-1} with the W47C spin label and 2.5 ± 0.7 s^{-1} with the D205C spin-label.

where c is the maximum relaxation rate for the 10-15 ns dipolar relaxation mechanism, too short to identify a discrete complex, but a reporter for membrane surface proximity, and the smaller magnitude cross-relaxation component. $\Delta R_{P\text{-}e}(0)$ is the maximum relaxation rate of the ^{31}P caused by the spin-label at the concentrations of each used and extrapolated to zero magnetic field; in this sample, the value for this spin-label affecting PC is $20 \pm 2\ s^{-1}$. The correlation time for the complex, $\tau_{P\text{-}e}$, is $1.4 \pm 0.2\ \mu s$. From these two values a distance between the spin-label unpaired electron and the ^{31}P of the phospholipid can be estimated with Equation 2.

$$r_{P\text{-}e}^{6} = ([BtPI\text{-}PLC]_o/\{(2/3)[phospholipid]_o)$$
$$x\ (S^2\ \tau_{P\text{-}e}/\Delta R_{P\text{-}e}(0))\ (0.3\ \mu^2(h/2\pi)^2\ \gamma_P^2\ \gamma_e^2) \qquad (2)$$

This assumes a single binding site for each phospholipid with a lifetime at least as long as the overall vesicle tumbling time. Also, since these are small, highly curved vesicles only about 2/3 of the phospholipid is in the outer leaflet (the distance of the spin-labeled protein for the inner leaflet is too far to have an effect on those phospholipid head groups). The key point is that the ratio $\tau_{P\text{-}e}/\Delta R_{P\text{-}e}(0)$ is proportional to $r_{P\text{-}e}^{6}$ where $r_{P\text{-}e}$ represents the distance between the ^{31}P and spin-label ($r_{P\text{-}e}$). This ensures that the PR_1E rapidly decreases as $r_{P\text{-}e}$ increases.

A similar analysis of ΔR_1 for PMe and spin-labeled W47C (Figure 6D) shows that the ^{31}P and spin-label on D205C are not very close to one another compared to D205C and PC. However, PMe is significantly closer to the spin-label on W47C. The longer $r_{P\text{-}e}$ values obtained for both these labels are consistent with PMe in the active site for the tumbling time of the vesicle. Comparing the effect of many different spin-labeled *Bt*PI-PLCs on PC provides approximate distances for each different ^{31}P-electron interaction [52]. Using Autodock to see where a PC analog (in this case butylphosphocholine) could bind and searching the protein surface for a phosphocholine site compatible with the NMR data, we identified a region on the periphery of the αβ-barrel that could specifically bind the phosphocholine moiety [52]. The complex most consistent with this NMR experiment was a PC choline cation – *Bt*PI-PLC tyrosine π complex (Figure 7). Although cation-π motifs are common in proteins, a similar interaction between a PC molecule in a membrane and a transiently bound protein had not been considered as a contributor to membrane-protein binding.

For *Bt*PI-PLC, the high-resolution field cycling method first showed that a specific complex for each lipid existed for at least about 1 μs. Obtaining $\Delta R_{P\text{-}e}(0)$ and $\tau_{P\text{-}e}$ for several mutants provided averaged ^{31}P-electron distances that, along with simple modeling, suggested where this novel PC binding site, a cation-π motif, might be. We have since used the methodology on several other amphitropic proteins listed below.

The isolated EF-hand domain of PLCδ1 binds in the absence of Ca^{2+} to PC membranes containing anionic phospholipids. By preferentially labeling each of two cysteine residues in this domain we could orient the protein on the membrane [53]. Once at the membrane surface it binds with nearly equal avidity to both anionic and PC lipids. That this domain can indeed bind phospholipids in this fashion suggested that when it is in the intact protein, its membrane binding aids in orienting the substrate in the active site for optimal catalysis.

The PH domain of Akt1 is translocated to the plasma membrane by binding to phosphatidylinositol 3,4,5-triphosphate ($PI(3,4,5)P_3$), whereupon the kinase is phosphorylated and activated. Alkylphospholipids such as perifosine bind to this domain (they also have cytotoxic properties), and it was originally thought that these neutral single chain phospholipids would bind to the phosphoinositide site. By spin-labeling this domain (which has two cysteines not in the immediate vicinity of the cationic phosphoinositide binding site), we could show that the perifosine binds near but not in the phosphoinositide site and is considerably closer to the spin-labeled cysteines [54]. A novel site was then identified by using the NMR distances as constraints and exploring the protein surface for a site that optimizes interaction of this zwitterionic lipid with the protein. Furthermore, when the perifosine binds to this site, the orientation of the alkyl chain would misalign the domain on the membrane. That in turn would prevent Akt phosphorylation and activation.

Lastly, PTEN, the key $PI(3,4,5)P_3$ phosphatase in the PI-3-kinase pathway, is activated by its product, phosphatidylinositol 4,5-bisphosphate ($PI(4,5)P_2$). Some mutagenesis results suggested that the N-terminal peptide of PTEN was involved in binding $PI(4,5)P_2$. However, the N-terminal peptide was missing from the crystal structure so this was difficult to model. By spin-labeling the protein and knowing that the PR_1E has a r^{-6} dependence on the distance between the ^{31}P and the unpaired electron of the spin-label, we could show that $PI(4,5)P_2$ binds in a site adjacent to the active site and in so doing will keep the protein anchored as substrates bind and products leave the active site [55].

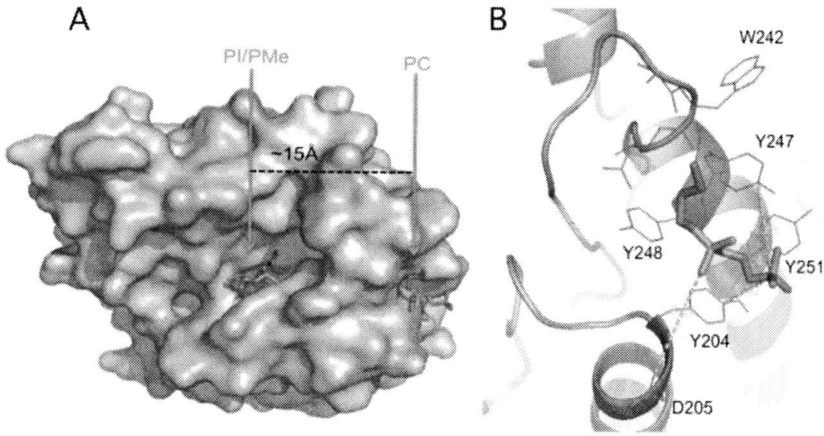

Figure 7. Autodock4 model for a butylphosphocholine ('PC') binding site on *Bt*PI-PLC. (A) Top view, looking down from the membrane at the spatial relationship of the novel PC binding site occupied by butylphosphocholine to the active site with a butylphosphoinositol bound (aligned with myo-inositol in crystal structure 1PTG). (B) Energy minimized structure around the activator binding site showing PC analogue interactions with nearby side-chains. This figure was originally published in Pu, M., Orr, A., Redfield, A. G., and Roberts, M. F. (2010) Defining specific lipid binding sites for membrane-bound phospholipase C, by ^{31}P NMR from 0.005 to 11.7 Tesla. *J. Biol. Chem.* 285:26916-22, as Figure 4 on p. 26920 © the American Society for Biochemistry and Molecular Biology.

In each of these cases, the amphitropic protein formed a specific complex with a phospholipid molecule at a site distinct from the substrate or natural ligand of the enzyme, which dramatically affected enzyme activity, in part by enhancing the partitioning of the protein onto a bilayer and/or facilitating substrate phospholipid access to the active site. We also use the observance of a low field PR_1E as the proof that a discrete phospholipid-protein complex must exist. Our upper limit for measuring correlation times indicates that if we see can measure a PR_1E, the complex must exist as least for that amount of time, 1-2 μs (and likely longer). In contrast, the 10 ns dipolar relaxation is short enough to reflect a nonspecific membrane proximity of the spin-label to the ^{31}P. We are currently developing a distance scale for a phospholipid head groups in a bilayer ^{31}P to spin-labels based on PR_1Es seen for this fast dipolar relaxation mechanism.

3.3. Computational Methods Exploring Amphitropic Protein Binding to Membranes

For the two experimental techniques described above, HDX-MS and high-resolution field cycling with spin-labeled proteins, a general membrane interface on the protein and in some cases specific phospholipid sites can be defined, as long as there is a structure for the protein in solution. However, these NMR or crystal structures solved in the absence of a lipid interface, are unlikely to capture any significant conformational changes adopted by the amphitropic protein when it is transiently anchored on the membrane surface. Computational approaches can provide in silico dynamics of the surface of the protein interacting with the membrane lipids and then assess the sorts of transient interactions that occur. Substrate [56] and inhibitor [57] binding to amphitropic enzymes can be particularly informative. Computational studies of two human phospholipase A_2 enzymes clearly showed that the membrane functions as an allosteric "ligand" that binds at the enzyme interfacial surface, and causes a conformational change from a closed (inactive) state in water to an open (active) state [56]. In a study looking at a series of dioctanoyl-3-deoxy-PI molecules binding to PTEN [57], the more potent D-series inhibitors adopted orientations on PTEN very similar to that of a $diC_8PI(3)P$ substrate. However, the L-series took advantage of an alternate hydrophobic pocket adjacent to the active site. Removing inositol hydroxyl groups weakened the affinity of the D- series compounds but enhanced the L-series binding and inhibiting PTEN.

MD simulations can be particularly insightful in characterizing novel lipid binding sites on proteins. For *Bt*PI-PLC binding to a PC bilayer, MD simulations tracked a number of PC cation – protein Tyr complexes over the course of the 500 ns simulation (Figure 8) [45]. These involved Tyr88, Tyr204, Tyr246, and Tyr251. However, the two most occupied sites involved Tyr88, closer to the active site, and Tyr246, on the barrel periphery.

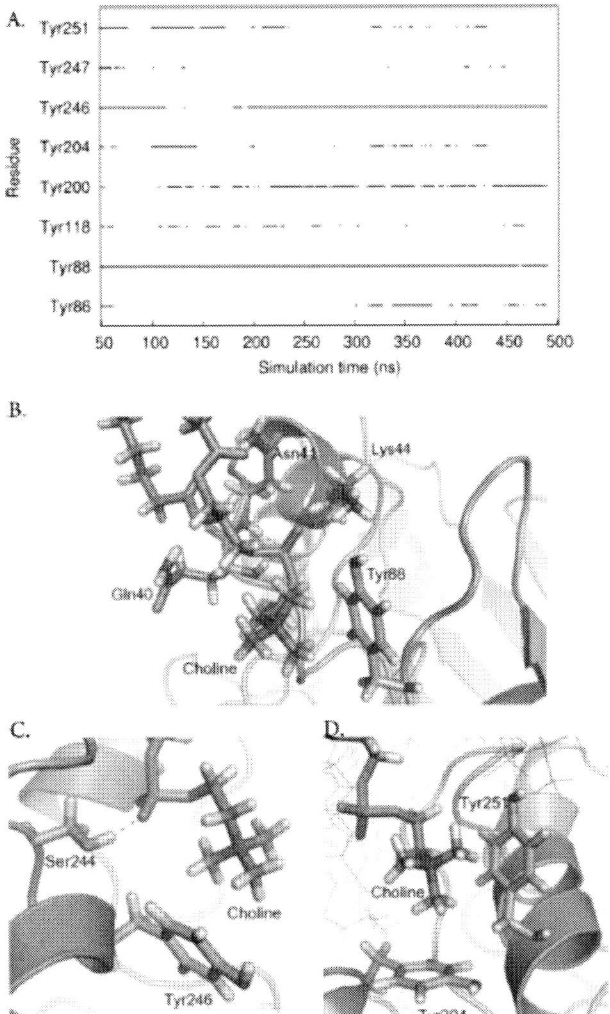

Figure 8. Cation-π interactions observed during MD simulations of the protein binding to a PC bilayer. (A) Occurrence of PC cation - Tyr-π interactions during the simulation. Note that complexes with Tyr88 and Tyr246 are the most persistent during the simulation. (B-D) Representative snapshots of cation-π interactions. This figure (in color) was originally published in Grauffel, C., Yang, B., He, B., Roberts, M. F., Gershenson, A., and Reuter, N. (2013) Cation-π interactions as lipid-specific anchors for phosphatidylinositol-specific phospholipase C. *J. Am. Chem. Soc.* 135:5740-50. Copyright 2013 the American Chemical Society.

When simulations (500 ns) were carried out with mixed bilayers of PG and PC up to 80 mol% PC, the strongest interaction was with Tyr246 [46]. This site agrees quite well with the experimental distances supplied by the field cycling experiments, which were carried out with equimolar PMe/PC SUVs [52]. Continuum electrostatics also showed another interesting facet of this protein. In general, it is assumed that amphitropic proteins are driven to membranes by electrostatics and a surface with several cationic residues and then anchored via hydrophobic contacts. Although the *Bt*PI-PLC electrostatic interactions vary with the bilayer composition, they are relatively weak and dominated by a single lysine residue [29]. Evaluating the distribution of basic amino acids in amphitropic protein structures then suggested that weak electrostatics may also be more common for peripheral membrane protein binding than thought. The computational analysis also shows that the cation-π interactions and not the hydrophobic ones vary with bilayer composition and thus correlate strongly with changes in apparent K_d for the enzyme binding to bilayers.

A review of this area of research with a case study of a yeast protein, Osh4, involved in lipid (cholesterol) homeostasis provides emphasizes the importance of the interplay between experimental and computational approaches [58].

4. Questions for the Future

While our knowledge of how amphitropic proteins bind to membranes has been dramatically enhanced, questions still remain. MD simulations play a major role in how we describe these interactions. However, there are some potential problems. For a more unusual mechanism like the PC cation – aromatic side chain π interaction, the force field that describes the interaction likely needs to be improved. Furthermore, most of the simulations show many transient interactions of protein side chains with membranes, but they are not long enough to monitor catalytic events with this type of enzyme and to sort out if significant conformational changes occur. For example, in the *Bt*PI-PLC simulations on PG/PC bilayers, a PG molecule is not usually found in the active site, yet our NMR studies indicate anionic phospholipids do bind to the active site when PC/PMe SUVs are used. Could this mean a conformational change promoted by the membrane is missing? Extending the time scale for simulations is likely to be extremely important for complex multidomain amphitropic proteins such as those in signal transduction pathways.

The emergence of PC choline cation – aromatic π complexes as a significant contributor to transient membrane binding for some proteins fits in with the idea that to control membrane localization of an amphitropic protein at a membrane surface, there should be multiple lipid anchors (which reduce lateral movement or partial dissociation from the membrane) that are not too tight. Tighter binding complexes of phosphoinositides with PH domains can be regulated by hydrolysis of the relatively rare phospholipid membrane anchor. What can regulate cation-π membrane interactions? For *Bt*PI-PLC, the cation-π motif serves as a way to target the enzyme to the PC-rich eukaryotic cell plasma membranes and no other bacterial membranes which will be negatively charged and lacking PC. To reduce the immune response, the enzyme has to release GPI-anchored proteins from many cells. Since cytoskeletal interactions and raft formation can constrain lateral diffusion of GPI-anchored proteins to small regions, or corrals, in the target membrane, the enzyme needs to bind transiently and promote hydrolysis of localized substrates, then to dissociate and either find another patch rich in GPI-anchors on the same cell or dissociate completely and bind to another cell. The time scale provided by the cation-π interactions should optimize these types of enzyme/GPI-anchor encounters.

Lastly, we now have many examples where specific phospholipids interact with an amphitropic protein to enhance membrane binding. Both rare and rather common phospholipids can form discrete complexes with these proteins. For an enzyme, such non-substrate interactions with a phospholipid at a site discrete from the active site classify the activating phospholipid as an allosteric effector, but with the caveat that perhaps more than a single phospholipid can be needed to trigger appropriate conformational changes, small or large, that occur with membrane binding. Treating individual lipids or a membrane patch as allosteric regulators of enzyme activity [56] as well as just enhancing anchoring to a membrane will require significant efforts, both experimental and computational, to describe the conformational details of many other amphitropic enzymes along the trajectory of catalysis.

REFERENCES

[1] Sudhahar, C. G., Haney, R. M., Xue, Y., and Stahelin, R. V. (2008) Cellular membranes and lipid-binding domains as attractive targets for drug development. *Curr. Drug Targets* 8:603-13.

[2] Moravcevic, K., Oxley, C. L., and Lemmon, M. A. (2012) Conditional peripheral membrane proteins: facing up to limited specificity. *Structure* 20:15-27.

[3] Hammond, G. R., and Balla, T. (2015) Polyphosphoinositide binding domains: key to inositol lipid biology. *Biochim. Biophys. Acta.* 1851:746-58.

[4] Lystad, A. H., and Simonsen, A. (2016) Phosphoinositide-binding proteins in autophagy. *FEBS Lett.* 590:2454-68.

[5] Scheffzek, K., and Welti, S. (2012) Pleckstrin homology (PH) domains – versatile modules in protein – protein interaction platforms. *FEBS Lett.* 586:2662-73.

[6] Vanhaesebroeck, B., Stephens, L., and Hawkins, P. (2012) PI3K signaling: the path to discovery and understanding. *Nat. Rev. Mol. Cell. Biol.* 13:195-203.

[7] Almena, M., and Mérida, I. (2011) Shaping up the membrane: diacylglycerol coordinates spatial orientation of signaling. *Trends Biochem. Sci.* 36:593-603.

[8] Corbalan-Garcia, S., and Gómez-Fernández, J. C. (2014) Signaling through C2 domains: more than one lipid target. *Biochim. Biophys. Acta.* 1838:1536-47.

[9] Mitchell, G., Ge, L., Huang, Q., Chen, C., Kianian, S., Roberts, M. F., Schekman, R., and Portnoy, D. A. (2015) Avoidance of autophagy mediated by PlcA or ActA is required for *Listeria monocytogenes* growth in macrophages. Infect. *Immun.* 83:2175-84.

[10] Sharom, F. J., and Lehto, M. T. (2002) Glycosylphosphatidylinositol-anchored proteins: structure, function, and cleavage by phosphatidylinositol-specific phospholipase C. *Biochem. Cell. Biol.* 80:535-49.

[11] Simon, M. C., and Kusch, J. (2013) Communicative functions of GPI-anchored surface proteins in unicellular eukaryotes. *Crit. Rev. Microbiol.* 39:70-8.

[12] Cho, W., and Stahelin, R. V. (2001) Membrane binding assays for peripheral proteins. *Anal. Biochem.* 296:153-61.

[13] Griffith, O.H., and Ryan, M. (1999) Bacterial phosphatidylinositol-specificphospholipase C: structure, function, and interaction with lipids. *Biochim. Biophys. Acta* 1441:237-54.

[14] Zhou, C., Wu, Y., and Roberts, M. F. (1997) Activation of phosphatidylinositol-specific phospholipase C towards inositol 1,2-(cyclic)-phosphate. *Biochemistry* 36:347-55.

[15] Zhou, C., Qian, X., and Roberts, M. F. (1997) Allosteric activation of phosphatidylinositol-specific phospholipase C: Phospholipid binding anchors the enzyme to the interface. *Biochemistry* 36:10089-97

[16] Qian, X., Zhou, C., Roberts, M. F. (1998) Phosphatidylcholine activation of bacterial phosphatidylinositol-specific phospholipase C towards PI vesicles. *Biochemistry* 37:6513-22.

[17] Wehbi, H., Feng, J., Kolbeck, J., Ananthanarayanan, B., Cho, W., and Roberts, M.F. (2003) Investigating the interfacial binding of bacterial phosphatidylinositol-specific phospholipase. *Biochemistry* 42:9374-9382.

[18] Fernandes, F., Coutinho, A., Prieto, M., and Loura, L. M. (2015) Eletrostatically driven lipid-protein interaction: answers from FRET. *Biochim. Biophys. Acta* 1848:1837-48.

[19] Bajar, B. T., Wang, E. S., Zhang, S., Lin, M. Z., and Chu, J. (2016) A guide to fluorescent protein FRET pairs. *Sensors* 18: E1488.

[20] Wong, J. K., Todd, M. H., and Rutledge, P. J. (2017) Recent advances in macrocyclic fluorescent probes for ion sensing. *Molecules 22*: 200.

[21] Mohsin, M., Ahmad, A., and Iqbal, M. (2015) FRET-based genetically-encoded sensors for quantitative monitoring of metabolites. *Biotechnol. Lett.* 37:1919-28.

[22] Krüger, D., Ebenhan, J., Werner, S., and Bacia, K. (2017) Measuring protein binding to vesicles by fluorescence cross-correlation spectroscopy. *Biophys.* J. 113:1311-25.
[23] Pu, M., Roberts, M. F., and Gershenson, A. (2009) Fluorescence correlation spectroscopy of phosphatidylinositol-specific phospholipase C monitors the interplay of substrate and activator lipid binding. *Biochemistry* 48:6835-45.
[24] Cheng, J., Karri, S., Grauffel, C., Reuter, N., Roberts, M. F., Wintrode, P. L., and Gershenson, A. (2013) Does changing the predicted dynamics of a phospholipase C alter activity and membrane binding? *Biophys. J.* 104:185-95.
[25] Feng, J., Webhi, H., and Roberts, M. F. (2002) Role of tryptophan residues in interfacial binding of phosphatidylinositol-specific phospholipase C. *J. Biol. Chem.* 277:19867-75.
[26] Feng, J., Bradley, W., and Roberts, M. F. (2003) Optimizing the interfacial activity of a bacterial phosphatidylinositol-specific phospholipase C. *J. Biol. Chem.* 278:24651-7.
[27] Shi, X., Shao, C., Zhang, X., Zambonelli, C., Redfield, A. G., Head, J. F., Seaton, B. A., and. Roberts, M. F. (2009) Modulation of *Bacillus thuringiensis* phosphatidylinositol-specific phospholipase C activity by mutations in the putative dimerization interface. *J. Biol. Chem.* 284:15607-18.
[28] Carman, G.M., Deems, R. A., and Dennis, E. A. (1995) Lipid signaling enzymes and surface dilution kinetics. *J. Biol. Chem.* 270:18711-4.
[29] Khan, H. M., He, T., Fuglebakk, E., Grauffel, C., Yang, B., Roberts, M. F., Gershenson, A., and Reuter, N. (2016) A role for weak electrostatic interactions in peripheral membrane protein binding. *Biophys. J.* 110:1367-78.
[30] Knight, J. D., Lerner, M. G., Marcano-Velazquez, J. G., Pastor, R. W., and Falke, J. J. (2010) Single molecule diffusion of membrane-bound proteins: window into lipid contacts and bilayer dynamics. *Biophys. J.* 99: 2879-87.

[31] Fox, C. B., Wayment, J. R., Myers, G. A., Endicott, S. K., and Harris, J. M. (2009) Single-molecule fluorescence imaging of peptide binding to supported lipid bilayers. *Anal. Chem.* 81:5130-8.

[32] Yang, B., Pu, M., Khan, H., Friedman, L., Reuter, N., Roberts, M. F., and Gershenson, A. (2015) Quantifying transient interactions between *Bacillus* phosphatidylinositol-specific phospholipase C and phosphatidylcholine-rich vesicles. *J. Am. Chem. Soc.* 137:14-7.

[33] Nath, A., Trexler, A. J., Koo, P., Miranker, A. D., Atkins, W. M., and Rhoades, E. (2010) Single-molecule fluorescence spectroscopy using phospholipid bilayer nanodiscs. *Meth. Enzymol.* 472:89-117.

[34] Barnaba, C., Martinez, M. J., Taylor, E., Barden, A. O., and Brozil, J. A. (2017) *J. Am. Chem. Soc.* 138:5420-30.

[35] Hohlbein, J., Heilemann, M., and Kapanidis, A. N. (2010) Surfing on a new wave of single-molecule fluorescence methods. *Phys. Biol.* 7:031001.

[36] Tran, R. J., Sly, K. L., and Conboy, J. C. (2017) Applications of surface second harmonic generation in biological sensing. *Annu. Rev. Anal. Chem.* 10:387-414.

[37] Tatulian, S. A. (2017) Interfacial enzymes: membrane binding, orientation, membrane insertion, and activity. *Methods Enzymol.* 583:197-230.

[38] Smirnova, T. I., and Smirnov, A. I. (2015) Peptide-membrane interactions by spin-labeling EPR. *Methods Enzymol.* 564:219-58.

[39] Vadas, O., Burke, J. E. (2015) Probing the dynamic regulation of peripheral membrane proteins using hydrogen deuterium exchange–MS (HDX-MS). *Biochem. Soc. Trans.* 43:773-86.

[40] Vadas O, Jenkins ML, Dornan GL, Burke JE. (2017) Using hydrogen deuterium-exchange mass spectrometry to examine protein-membrane interactions. *Methods Enzymol.* 583:143-72.

[41] Heinz, D. W., Ryan, M., Bullock, T. L., and Griffith, O. H. (1995) Crystal structure of the phosphatidylinositol-specific phospholipase C from *Bacillus cereus* in complex with *myo*-inositol. *EMBO J.* 14:3855-63.

[42] Koshy, C., and Ziegler, C. (2015) Structural insights into functional lipid-protein interactions in secondary transpoters. *Biochim. Biophys. Acta* 1850:476-67.

[43] Cheng, J., Goldstein, R., Gershenson, A., Stec, B., and Roberts, M. F. (2013) The cation box is a specific phosphatidylcholine membrane targeting motif. *J. Biol. Chem.* 288:14863-73.

[44] Goh, B. C., Wu, H., Rynkiewicz, M. J., Schulten, K., Seaton, B. A., and McCormack, F. X. (2016) Elucidation of lipid binding sites on lung surfactant protein A using x-ray crystallography, mutagenesis, and molecular dynamics simulations. *Biochemistry* 55:3692-701.

[45] Grauffel, C., Yang, B., He, B., Roberts, M. F., Gershenson, A., and Reuter, N. (2013) Cation-π interactions as lipid-specific anchors for phosphatidylinositol-specific phospholipase C. *J. Am. Chem. Soc.* 135:5740-50.

[46] Redfield, A. G. (2012) High-resolution NMR field-cycling device for full-range relaxation and structural studies of biopolymers on a shared commercial instrument. *J. Biolmol. NMR* 52:159-77.

[47] Roberts, M. F., and Redfield, A. G. (2004) High resolution ^{31}P field cycling as a probe of phospholipid dynamics. *J. Amer. Chem. Soc.* 126:13765-77.

[48] Roberts, M. F., and Redfield, A. G. (2004) Phospholipid bilayer surface configuration probed quantitatively by ^{31}P field-cycling NMR. *Proc. Natl. Acad. Sci. U.S.A.* 101:17066-71.

[49] Klauda, J. B., Roberts, M. F., Redfield, A. G., Brooks, B. R., and Pastor, R. W. (2008) Rotation of lipids in membranes: MD simulation, ^{31}P spin-lattice relaxation, and rigid-body dynamics. *Biophys. J.* 94:3074-83.

[50] Roberts, M. F., Mohanty, U., and Redfield, A. G. (2009) Phospholipid reorientation at the lipid/water interface measured by high resolution field cycling ^{31}P NMR spectroscopy. *Biophys. J.* 97:132-41.

[51] Zhang, X., Wehbi, H., and Roberts, M. F. (2004) Crosslinking phosphatidylinositol-specific phospholipase C traps two activating

phosphatidylcholine molecules on the enzyme. *J. Biol. Chem.* 279: 20490-500.

[52] Pu, M., Orr, A., Redfield, A. G., and Roberts, M. F. (2010) Defining specific lipid binding sites for membrane-bound phospholipase C, by ^{31}P NMR from 0.005 to 11.7 Tesla. *J. Biol. Chem.* 285:26916-22.

[53] Cai, J., Guo, S., Lomasney, J. W., and Roberts, M. F. (2013) Ca^{2+}-independent binding of anionic phospholipids by phospholipase C δ1 EF-hand domain. *J. Biol. Chem.* 288:37277-88.

[54] Gradziel, C. S., Wang, Y., Stec, B., Redfield, A. G., and Roberts, M. F. (2014) Cytotoxic amphiphiles and phosphoinositides bind to two discrete sites on the Akt1 PH domain. *Biochemistry* 53:462-72.

[55] Wei, Y., Stec, B., Redfield, A. G., Weerapana, E., and Roberts, M. F. (2015) Phospholipid binding sites of PTEN: Exploring the mechanism of PIP_2 activation. *J. Biol. Chem.* 290:1592-606.

[56] Mouchlis, V. D., Bucher, D., McCammon, J. A., and Dennis, E. A. (2015) Membranes serve as allosteric activators of phospholipase A_2, enabling it to extract, bind, and hydrolyze phospholipid substrates. *Proc. Natl. Acad. Sci.* U.S.A. 112:E516-25.

[57] Wang, Q., Wei, Y., Roberts, M. F., and Krilov, G. (2010) Understanding the stereospecific interactions of 3-deoxyphosphatidylinositol derivatives with the PTEN phosphatase domain. *J. Molec. Graphics Model.* 29:102-14.

[58] Monje-Galvan, V., and Klauda, J. B. (2016) Peripheral membrane proteins: tying the knot between experiment and computation. *Biochim. Biophys. Acta* 1858:1584-93.

In: Protein-Lipid Interactions
Editor: Angel Catalá
ISBN: 978-1-53613-125-3

Chapter 4

Lipid-Dependent Gating of Ion Channels

*Qiu-Xing Jiang**
Department of Microbiology and Cell Science,
University of Florida, Gainesville, FL, US

Abstract

Ion channels are important because all live cells rely on them to establish a transmembrane electrostatic potential and ion gradients across their membranes. They all are integrated inside lipid bilayers and form ion conducting pores whose open states allow passive flow of certain ions down their electrochemical gradients. The dissipative flow of ions usually sends a signal at the cost of cell's energy. In general, ion channels face two aqueous phases and a lipid phase. Via chemical interactions through physical contact, ion channels can sense changes in all three phases. Ion channels are good model systems for studying lipid-channel interactions because patch-clamp recordings of single or multiple channels can be made even though manipulation of lipids immediately around ion channels is difficult to achieve. Based on the effective distance of chemical forces and the physical size of lipids, the first layer of lipid molecules, called the annular lipids, next to the transmembrane domain of

* Corresponding Author Email: qxjiang@ufl.edu or jiangq9992003@gmail.com.

a channel dominates the lipid-protein interaction. We propose that the annular lipids and the channel form a functional unit and the former need to rearrange in company with the allosteric conformational change in the latter. The exchange of lipids between the annular layer and the bulk lipid phase is therefore an important factor. In consideration of the chemical interactions, there are specific interaction sites for specific lipid molecules on proteins through induced-fit, which have sub-mM or higher affinity and could be stereo-specific. Low affinity binding sites are often nonspecific and may be less sensitive to the stereo-chemistry of the lipids. Lipid-dependent gating of voltage-gated ion channels refers to channel's sensitivity to the chemical differences between phospholipids and nonphospholipids. The presence or absence of phosphate groups causes a major chemical difference in charge density and H-bonding network such that the energetic differences of a channel protein between the open and closed states may become sufficiently significant in the two groups of lipids. In this chapter, we will focus on the prior studies that demonstrated significant lipid-dependent gating effects on ion channels, and present a systematic thermodynamic description, which may be suitable to explaining many, if not all, of the prior observations. We will discuss the current technical difficulties in studying the lipid-dependent gating and possible solutions in addressing them in the future.

1. INTRODUCTION

Lipid environments form a significant electrochemical barrier for cells to separate themselves from their environments and establish an off-equilibrium state so that all activities of life may happen [1]. All biological lipids are amphipathic, containing both water-like, hydrophilic heads and water-dislike, hydrophobic tails. Almost all biological membranes, except the known membrane enclosure for lipid droplets, are made of two layers of lipids molecules that arrange themselves to have a hydrophobic core and two water-facing surfaces. These two layers (leaflets) are mostly separated, and are cross-linked in certain organisms living in extreme environments. Without chemical linkage, the two leaflets attached to each other through hydrophobic interactions, squeezing water molecules out of the middle of bilayers such that the water concentration in the hydrophobic core is 2-3 orders of magnitude lower than in the aqueous phase. The hydrophobic core is therefore self-driven to form a continuous membrane unless it is stabilized in solution by other amphipathic molecules at the water-facing edges. It constitutes a unique chemical environment for membrane proteins.

Membrane proteins are major chemical responders in cell membranes that are responsible for orchestrating a structural support of membrane and for sending information and materials across the bilayer barrier [1]. Based on their functions, membrane proteins can serve structural roles, transmit allosteric information across membrane, or pump/pass specific molecules or solutes from one side of the membrane to the other. All canonical ion channels are integral membrane proteins [2]. There are also well-known cases that certain soluble proteins may change conformation and become inserted into lipid bilayers and form ion channels that are either specific or nonspecific in allowing certain solutes to pass their conducting pores [3-7]. Except these disruptive channels, nearly all channels have at least two different states --- open and closed. The open state refers to the conformation that the conducting pathway allows specific solutes to pass, and the closed state is the one that does not. A controlled switch between the open and closed states of an ion channel is called the "gating" of the channel. Almost all ion channels are tightly gated so that they do not stay open and continuously dissipate off-equilibrium chemical potential difference that is established by active, energy-consuming processes across cell membranes.

Lipid membranes provide an environment to stabilize ion channels in them and affect the energetic status of the channel proteins. It is generally believed that lipids in membranes must have specific and nonspecific interactions with membrane proteins. The past studies have showed that certain ion channels have specific requirements for phospholipids for their functions. PIP_2 is known to regulate the function of IRKs, Kv channels as well as other membrane proteins [8-15]. Negatively charged lipids are required for proper folding and function of KcsA [16]. Anionic lipids are important for the MscL channel [17]. Polyunsaturated fatty acid and lyso-lecithin have been found to modulate ion channels [18-26]. Cholesterol and sphingolipids are major components in many cells, especially in neurons and glial cells [27-34]. Cholesterol-binding pockets have been resolved in bovine rhodopsin [35], nicotinic acetylcholine receptor (nAchR) [36-43], GABA receptors [44, 45], serotonin receptors [39], IRK channels, etc.

[29, 46, 47] It has been taciturnly assumed by many researchers that the specific interactions through high-affinity binding pockets are the dominating factor, and therefore most of the published studies have tried to define the residues lining the binding pockets for lipid molecules in different membrane proteins. It is interesting to note that in lipid membranes, the local concentrations of lipids may be quite high (hundreds of milli-molar) and the lateral movement of lipids in a fluidic phase across a distance of one micrometer only takes about a few milliseconds, suggesting that the high-affinity binding pockets are very likely saturated when the targeting lipids are relatively abundant (more than 1 mol%). Such binding pockets would be emptied if the targeting lipids are cycled through different endomembranes or converted into other intermediate forms, such as PIP_2 and other inositol phosphates. There have been multiple reviews on the specific lipid-binding pockets for different lipids in different ion channels [29, 48-50]. The readers may consult them for further details.

For cholesterol, whose membrane concentration may vary in a range of 5-45 mol% in different areas of the cell membranes, the effective binding of cholesterol in a specific binding pocket of an ion channel may be saturated such that the effects of cholesterol could only be detected when chemical depletion or chemical loading changes the cholesterol content out of the normal physiological range [32-34, 51, 52]. There are other changes in cell structure and cell function when chemical methods are used to alter cholesterol content, which are therefore expected to cause significant variations in the observed effects of cholesterol manipulation on voltage-gated channels. Published data on Kv, Cav and Nav channels support this notion, suggesting a difficult scenario in the physiological significance of cholesterol-dependent gating effects on ion channels, which is an integral part of the lipid-dependent gating discussed below.

The nonspecific interaction between lipids and membrane proteins can happen at multiple sites where the annular lipids are physically in contact with the proteins. The binding affinity could be at a mM-μM level such that the lipids are not tightly bound and usually are missing in purified membrane proteins in detergents. However, when the contact sites are in a group and concerted in their binding by the targeting lipids, the collective thermodynamic effects might be substantially high such that multiple nonspecific binding sites together may introduce significant changes in the gating of ion channels, and more broadly in the function of other membrane proteins. The lipid-dependent gating of ion channels is more in this category and focuses on the differences between phospholipids and nonphospholipids, and will be the main focus of our discussion.

In this chapter, we will first discuss the basic properties of lipid-dependent gating of Kv channels and some fundamental properties of native lipids in cell membranes that are pertaining to lipid-dependent gating [51], and then introduce the general concept of collective interaction between annular lipids and the transmembrane domain of a protein. A biophysical model is introduced to account for the nonspecific contacts between annular lipids and the channel, and a simple thermodynamic analysis is introduced to analyze the energetic principles behind the nonspecific interaction sites. A comparison with the annular lipids around Ca-ATPase in crystal structures will be introduced to strengthen the concept that in a similar fashion, the annular lipids are able to rearrange themselves in company with the allosteric conformational changes of the channel [53]. After that, we will discuss some controversial results from molecular dynamics simulation and a solid-state NMR study. To emphasize the importance of allosteric changes in the annular lipids, we will compare lipid distribution around aquaporin-0 and a Kv channel as specific examples for static lipid distribution and dynamic lipid movements, respectively. As a physiological example highlighting the importance of the lipid-dependent gating, a brief survey will be introduced to demonstrate the severe neurological effects of lipid metabolic defects and the functional defects at the systems level and the organ structural level. Lastly, we will discuss the current major technical limitations in studying quantitatively lipid-protein interactions with sufficient resolutions in time and space and propose possible solutions for future development.

2. What Is Lipid-Dependent Gating?

The concept of lipid-dependent gating started with the question on the possible difference between phospholipids and nonphospholipids in their interactions with the voltage sensor domains (VSDs) of a Kv channel [51, 54-56]. Our studies of Kv channels in homogeneous lipid bilayer membranes recognized that synthetic nonphospholipids stabilize the KvAP channel in the closed state with its VSDs in the resting conformation, and the lipid-stabilized resting state is equivalent to the native resting state driven by hyperpolarization [51]. These results led to our working hypothesis of "lipid-dependent gating" (Figure 1) [57]. This new phenomenon reflects the effects of the annular lipids right next to the channels (especially the VSDs) that cause a static change in the energetic levels of two (or more) distinct states. Such a change in annular lipid distribution is slower than the voltage-dependent gating. This hypothesis explains the gating charge immobilization caused by sphingomyelinase treatment of cell membranes [58] and may play a role in the down-regulation of activities of Kv clusters in lipid domains in cell membranes [59]. Similar nonphospholipid-driven conformational changes have been reported for hyperpolarization-activated MVP channels in a homogeneous membrane system where conclusive data were collected to tie a strong connection between the structural changes of the channel and the altered lipid environment [60]. These data not only reproduced our observations made with the KvAP channels as well as those from eukaryotic Kv channels in cell membranes treated with different sphingomyelinases, but also extended the scope of our working hypothesis and raised a high likelihood that lipid-dependent gating be a generic mechanism underlying the inhibitory effects of nonphospholipids on voltage-gated ion channels.

Physically the structural basis of the conformational changes induced by the nonphospholipids must result from the direct contacts between the lipids and the VSDs of the Kv channels. So far the lipid-dependent gating effect has been observed in multiple Kv channels. It could be less prominent for voltage-gated sodium (Nav) or calcium (Cav) channels [30, 32, 34, 61-73] because the latter contain single polypeptides with 4 different VSDs playing differential roles in voltage-dependent gating of the ion-conducting pores, and their VSDs are usually shielded (at least partially) from annular lipids by auxiliary transmembrane subunits [74-77]. The shielding would decrease the number of contact points between the annular lipids and the VSDs, and mitigate the energetic changes caused by the lipid-VSD interactions. Due to the difficulty in producing large quantities of recombinant Nav or Cav channels, direct measurements of lipid-dependent gating of the Nav or Cav in the absence or presence of their auxiliary subunits are not done yet and will be required to test the generic applicability of lipid-dependent gating to voltage-gated ion channels.

3. Lipid Properties in Membrane That Are Important for Lipid-Dependent Gating

In human cell membranes, all lipid molecules belong to two general groups (Figure 2): phospholipids (Group I) and nonphospholipids (Group II), depending on whether there are phosphate groups in the headgroup regions. The phospholipids have been well known for a long time and have been used in many reconstituted membrane systems. The main chemical difference between phospholipids and nonphospholipids is that the negatively charged phosphate group is a capable of forming a hydrogen-bonding network with water molecules in between and functional groups of proteins such that the arrangement of the phosphate groups in the annular lipids at the similar height around the transmembrane domain of a membrane protein constitutes a "phosphate-belt", and stabilizes the membrane protein in specific conformations. A good example has been observed by electron crystallography of aquaporin-0 in DMPC membrane [78, 79].

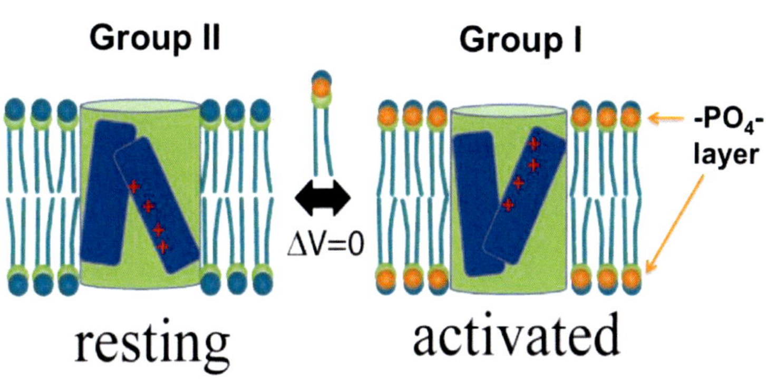

Figure 1. Schematic lipid-dependent gating. Group I lipids refer to phospholipids including both glycerophospholipid and sphingomyelin. Group II lipids include all lipid molecules that do not have phosphate groups in their headgroup regions. A VSD (blue bars with + charges in red) next to a pore domain (green cylinder) switches between the resting ("down") and the activated ("up") state in Group II and Group I lipids at $\Delta V = 0$.

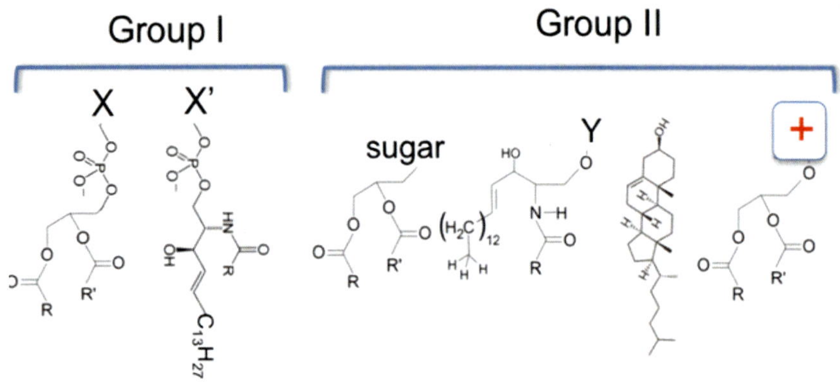

Figure 2. Phospholipids and nonphospholipids in eukaryotic cell membranes. Group I includes glycerophospholipids and sphingomyelin. The X group usually is –H, choline, serine, ethanolamine, glycerol, or phosphoinositol [including the glycophoshoinositol (GPI)]. The X' group is mostly choline. The group II is more diversified and can be glycerol glycolipids, sphingolipids (excluding sphingomyelins; Y is not a choline phosphate), cholesterol and positively charged glycerol lipids. In archae, the cell membranes usually contain a large fraction of ether-lipids without phosphate groups, and some of them are cross-linked at the ends of the hydrophobic tail so that the two leaflets in the bilayer are covalently bonded for high thermal stability and tight packing of lipids to maintain the physical barrier against transmembrane permeation.

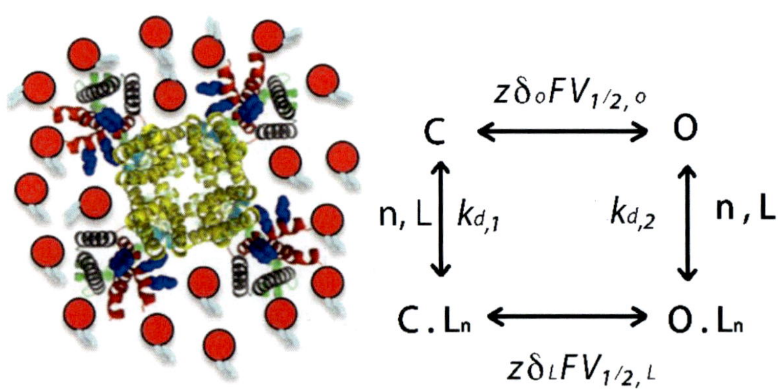

Figure 3. A working model for the lipid-dependent gating effects on a Kv channel. Left: Annular lipids (pinkish red circles represent the headgroups) around a channel exert most effects because of their physical contact with the voltage-sensor domains (VSDs). Lipids between two VSDs may affect the allosteric change of the VSDs. So are the lipids on the peripheral side of the VSDs. Based on the average dimension of lipid molecules and that of a typical Kv channel, each channel has ~60 lipid molecules in the annular layer in each leaflet of the bilayer membrane. Right: a simple n-site binding model for a two-state gating model. To account for changes in the apparent gating charge, at least one more closed state should be added.

For many of us, membrane lipids are usually limited to the glycerol phospholipids, not including even sphingomyelin. Cholesterol is well known. Other members of the Group II lipids are quite unfamiliar because of limited understandings of their physiological roles in cell membranes. Group II includes glycosphingolipids, glyceroglycolipids [excluding glycophosphotidylinositols (GPI)], cholesterol and its analogs, and the positively charged cation lipids. The polyunsaturated fatty acids are special because of the multiple different interaction sites that have been proposed when they were applied to the channels in membranes [21]. Genetic evidence from patients afflicted with metabolic defects in Group II lipids suggests that they are very important for the normal function. Group II lipids are believed to be involved in maintaining the structural stability of cell membranes, cell identity due to surface modifications on Group II lipids and lipid asymmetry, domain organization of lipids in membranes, and signaling events due to specific lipid-protein interaction. From a general standing point, the essential role and the physiochemical basis of lipid organization and phase separation remain poorly understood, especially the functional effects of lipid domains on specific membrane proteins. The main cause of such knowledge gaps is technical limitations in analyzing lipid composition in small membrane patches.

From an evolutionary standpoint, membrane proteins are evolved to function in complicated lipid environments. The inhibitory effect of Group II lipids on Kv channels predicted by the lipid-dependent gating hypothesis is a manifestation of the protein adaptation to lipidic environments in membranes. Defects in such adaptation are expected to cause severe defects in different types of eukaryotic cells, and in cell-to-cell communications. From this angle, lipid-dependent gating may be extended to other membrane proteins, such as GPCRs, cytokine receptors, etc.

4. A Biophysical Model for Lipid-Dependent Gating of Kv Channels

To provide a biophysical model for lipid-dependent gating, we can start with a general model (left panel in Figure 3), assuming that annular lipids interact in low, but different affinities with a Kv channel in the resting (C) and activated (O) states (compared with Figure 1). A structural model of the KvAP channel, whose pore region is depicted in yellow and the VSDs in grey, was generated from homologous modeling based on the Kv1.2 structure [80], and is immersed in a bilayer membrane. The pinkish red circles with white tails represent individual lipid molecules in the annular layer. We are not considering the crevices in the protein that are occupied by specific lipids with higher affinity. The bulk phase lipids beyond the annular layer are neglected here because short-range interactions dominate the physicochemical forces underlying the lipid-dependent gating effect. In response to change in transmembrane electrostatic potential, the annular lipids will move in harmony with the VSDs switching between different gating conformations. Protein binding may also induce a shift in statistical distribution of specific lipids between the annulus and the bulk phase in membranes. A shift in gating transition, which refers to the switching between the resting and activated states of a channel and is represented by conductance-voltage relation (*G-V*) for the channel pore and the gating charge-voltage (*Q-V*) relation for the VSDs, reflects differences in apparent binding affinity of the annular lipids to channels in two distinct states. If we assume no change in apparent gating charge, using δ_0 and δ_L for the apparent charge for the *G-V* relations measured from ligand-free (**C** and **O**) and –bound (**C.L$_n$** and **O.L$_n$**) channels, respectively, the measured *G-V* curves can provide an estimated change in free energy between the activated and the resting states (right, Figure 3).

$$\Delta\Delta G = z\delta_L F V_{\frac{1}{2},L} - z\delta_0 F V_{\frac{1}{2},0} = -< n > RT ln\left(\frac{k_{d,1}}{k_{d,2}}\right) \quad (1)$$

Here, $<n>$ is the average number of a specific lipid of interest (a nonphospholipid as a ligand, *L*) in the annulus per channel, varying from 0 to 120 depending its molar ratio in the annular layer, *[L]* apparent concentration of the target lipid, *F* Faraday constant, *RT* gas constant times temperature, $z\delta$ apparent gating charge, $k_{d,1}$ and $k_{d,2}$ apparent dissociation constants assuming equivalent low-affinity binding sites without mutual interference, $RT\ ln\ (k_{d,1}/k_{d,2})$ is approximately the unitary energy difference caused by each ligand-binding event to two states, and $V_{1/2}$*'s* the mid-points of *G-V* curves. As a simplified case, $k_{d,1}$ & $k_{d,2}$ are assumed to be approximately the same for a maximal total of *N* binding sites per channel in two gating conformations. At low *[L]*, <*n*> becomes unity. *G-V* curves will be right-shifted when a specific ligand binding event stabilizes the closed state ($k_{d,1} < k_{d,2}$). With proper experimental conditions, it might even be feasible to record gating-charge movement in the VSDs (giving rise to the gating currents) and define the gating charge-voltage relation (*Q-V*) for ligand-free and –bound states, and use the shift of *Q-V* curves to estimate the lipid-dependent effects on VSD movement. A *ΔΔG-[L]* plot can be experimentally measured. When *[L]* is very low (dilute), ΔΔG approaches $-RT\ ln\ (k_{d,1}/k_{d,2})$. At saturating *[L]*, ΔΔG ~ - $NRT\ ln\ (k_{d,1}/k_{d,2})$. These two conditions thus provide a low-bound estimate of the total number of binding sites, *N*. The average dimensions of a tetrameric channel and lipid molecules suggest ~120 contact sites for lipids in the annulus of each channel, ~60 in each leaflet. Calculated *ΔΔG* based on shifting the *G-V* curve of a typical *Shaker*-like Kv channel (~3-4 apparent elementary charges) by 200 mV without changing its steepness ($\delta_0 \sim \delta_L$) is ~15 kcal/mol. The actual number of total gating charge per channel from the movement of four VSDs may vary from 10-13.5 elementary charges. On a practical scale, a significant change can be contributed by ~10 mol% nonphospholipids among the annular lipids; Each ligand-binding site only contributes a small energetic change of ~1-2 *RT*, which means merely a 3-8 fold change in the apparent binding affinity ($k_{d,1}/k_{d,2}$) at each contact site. It suggests that collectively a small change in the low-affinity binding sites for multiple ligand-binding sites is sufficient to cause a major shift in the *G-V* and a significant inhibitory effect on the channel function. The same is likely true for the *Q-V*.

The above simplified model assumes that lipids in membranes are well mixed such that the bulk phase and the annulus have the same lipid composition. The fluidity of the lipids can be monitored by FRAP (fluorescence recovery after photobleaching) or imaged using fluorescent lipids that are sensitive to macroscopic phase separation (81). But even with super-resolution fluorescence microscopy, we are still not able to observe lipid distributions in the annular layer or the bulk phase. Due to high concentrations (~50-500 mM for 5-50 mol%) of major nonphospholipids in eukaryotic cell membranes, all high-affinity binding sites (crevices at the hydrophobic surface of a channel) must be fully saturated. It is thus mainly low affinity binding sites that contribute to the measured change in the Gibbs free energy, *ΔΔ*G.

A major advantage of the above theoretical framework is that the effects of individual lipid binding events can be compared by $\Delta\Delta G$ in different membranes. It suggests that we may use binding property of individual lipids to predict the lipid-dependent gating effects in composite membranes. Relative efficacy of nonphospholipids in regulating a Kv channel will be comparable when lipid composition is changed and the effect of individual lipids is quantitatively measured. Additivity and/or mutual substitution of the gating effects of different lipids can be assessed. If successfully, such quantitative analysis can be extended to more complicated gating models containing more gating states than the two-state model used in our current analysis.

As an example, we analyzed the *G-V* relations calculated from the published data obtained from KvAP channels in cholesterol-containing bSUMs (Figure 4) [54]. The KvAP channels were purified and inserted in DOPC membranes containing 5.0% weight ratio (or 9.6% molar ratio) cholesterol in a special bead-support unilamellar membrane (bSUM) which favors the outside-out insertion of the channel and suffers no effect of the solvent used in the black lipid membranes. The bSUMs are fluidic and easy to control the lipid composition of cholesterol : DOPC mixture during membrane reconstitution. Consistent with the prediction from the biophysical model (Figure 3) and earlier observations [51], 9.6 mol% cholesterol caused a significant inhibitory effect on the KvAP opening. In such membranes, KvAP only started to show activity when the transmembrane voltage was above +60 mV [54]. Biophysical analysis shows that 9.6 mol% CHOL in DOPC membrane still keeps high membrane fluidity [81]. Compared to channels in the POPE/POPG (PEPG) membrane, the *G-V* curve of KvAP channels in 9.6 mol% CHOL-containing DOPC membranes was right-shifted by ~210 mV. Energetically, CHOL stabilized the channels in the closed state by ~15 kcal/mol. The free energy change $(\Delta\Delta G)$ could only be marginally accounted for by changes in ON- and OFF-rates (~3.3 kcal/mol; data not shown), suggesting that the stabilization effect is probably dominated by the CHOL effect on the VSD movement, which differs from the proposed replacement of phospholipids bound at the S4/S5 linkers to affect the last opening step of the channel (50, 73). In a homogeneous membrane, the CHOL effect is much stronger than what was reported in chemically treated cells [32, 34], which were probably complicated by the presence of

CHOL-free domains, changes in membrane protein distribution, vesicle fusion or endocytosis. Without these complications, our analysis in well-controlled bSUMs is much more accurate, quantitative and reproducible.

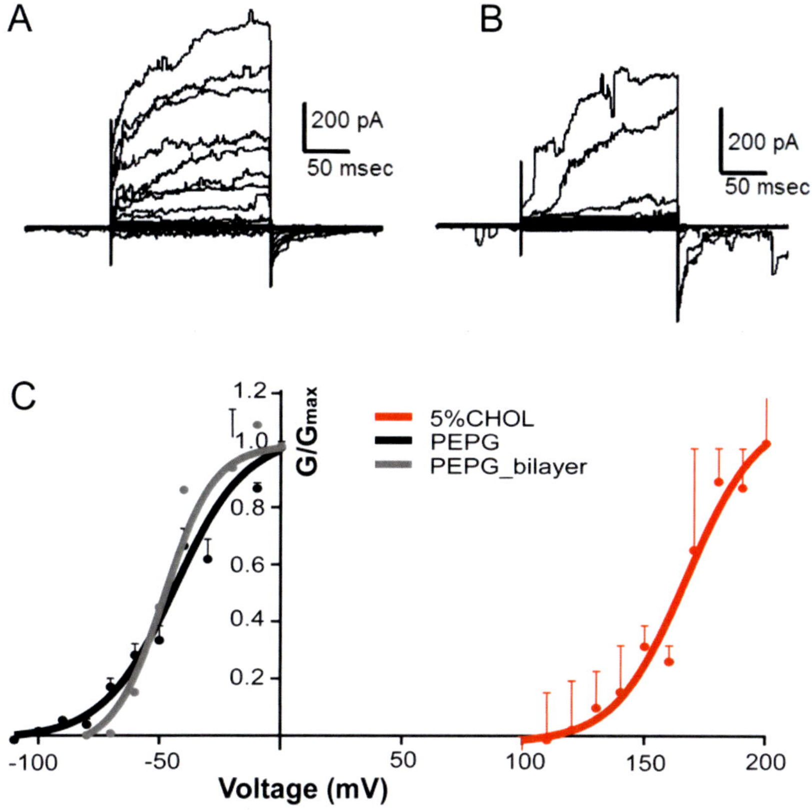

Figure 4. CHOL inhibits KvAP activity in a homogeneous membrane. A) Activity of KvAP in bSUMs made of POPE:POPG (PEPG, 3:1 weight ratio), pulsed from -160 to 180 mV at 10 mV steps. B) Recordings from KvAP channels in a bead-supported unilamellar membrane (bSUM) made of DOPC plus 9.6 mol% of cholesterol (5% CHOL in weight ratio). C) Normalized *G-V* for KvAP in bSUMs made of PE:PG (3:1, black; n = 4), DOPC/CHOL (9.6 mol% CHOL in brown; n = 3) and as a control in bilayer membranes made of POPE: POPG = 3:1 (weight ratio, PEPG_bilayer in grey; n = 3). Boltzmann fittings (solid lines) yielded (□ δ, $V_{1/2}$) (e_0, mV) for channels in PEPG_bilayer (3.0, -47.5), PEPG bSUMs (2.0,-45), and 9.6 mol% CHOL (2.0,162) in DOPC bSUMs. Error bars: s.d.

The simple model proposed above may be applied to explaining the cholesterol-dependent gating effect on the IRK channels and the pLGICs, where the binding sites have higher affinity and thus the effects from a smaller number of sites become sufficiently significant.

5. Annular Lipids and the High-Affinity and Low-Affinity binding sites Around Ca-ATPase

The readers might be skeptical about the proposed annular lipids in the proposed simple model and whether it is physically feasible despite apparent simplicity. Recently, the Ca-ATPase has been crystallized in stacked detergent-saturated membranes in different states of its catalytic cycle when contrasting reagents were used to enhance the resolution of individual lipid molecules in the annular layer (Figure 5) [53]. These results support the idea that the annular lipids may act on both high-affinity and low-affinity binding sites at the surface of the transmembrane domains and they are rearranged in order to adapt to the allosteric conformational changes in the transmembrane domains.

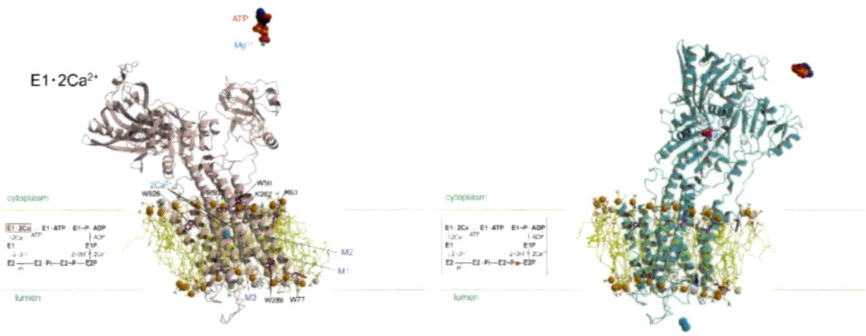

Figure 5. Annular lipids move in company with the allosteric motion of Ca^{2+}-ATPase. Left: Lipid molecules (red headgroups and yellow acyl tails) attached to the transmembrane domain are tilted to the left side when the ATPase is in the E1-2Ca state before the ATP binding. Right: Lipid molecules rearrange themselves in accord with the rearrangement of the transmembrane helices when the ATPase changes to the E2-P state after Ca^{2+} (Cyan balls) is released to the luminal side. The allosteric changes in both lipids and the transmembrane domains suggest that the lipid-protein interactions at both specific and nonspecific binding sites play a critical role in the reaction cycle of the ATPase.

Relating to the lipid-dependent gating hypothesis, the two main points from crystallographic studies agree well with what we discussed in the biophysical model (Figure 3), and are expected to hold true for other membrane proteins that undergo significant conformational changes among their key functional states. In particular, for the Kv channels, our model predicts that the annular lipids rearrange following the voltage-driven conformational changes of the VSDs. The tight packing of the nonphospholipids would change the energetic states of the channel and shift the equilibrium between the resting and the activated states. For the channels that do not undergo major conformational changes, such as the aquaporin-0 in the open state [78], the annular lipids seen by electron crystallography would probably be pretty static and not need to go through obvious rearrangements such that the lipid-dependent gating effect from the nonspecific binding is expected to be pretty weak, even though we cannot exclude the possible requirements of specific lipids that must be present for channel stability in membrane or for its water conduction. Combined all these observations and the theoretical models, we propose that the nonspecific lipid-channel interaction may become a major contributor to the energetic change of the channel in membrane, which is a relative level and may be shifted to favor the channel in either the closed or the open state. In the nonphospholipid-dependent inhibitory effects observed by us and other research groups on different Kv channels, the VSDs happen to be favored in the resting state by nonphospholipids as we have measured. The generic role of nonphospholipids on voltage-gated ion channels will need further experimental investigations.

6. Revisit Some Results from Molecular Dynamics Simulation and Solid-State NMR Studies

When coarse-grain molecular dynamics simulation was used to compare S4 helix-membrane interaction when the KvAP S4 voltage-sensor helix was changed from a POPC bilayer and a bilayer of DOTAP [82], a synthetic nonphospholipid, no significant conformational change was observed in a period of 60 ns. The negative counter ions were found to replace the anionic phosphate groups and stabilize the arginine residues. A local reorientation of the gating-charge contributing residues and deeper penetration of aqueous crevices around the S4 were proposed to account for the stability of the S4 (immobilization) in a more water-permeable

DOTAP bilayer and thus more resistance to voltage-driven conformational change. This explanation is inconsistent with the major allosteric changes of the VSDs observed by experimental measurement. It could either result from short MD simulation that did no allow the changes to happen or be due to the limited accuracy in the calculations of the energetic changes needed for the conformational changes of the isolated S4 helix.

When the isolated KvAP VSD was simulated in a PC bilayer [83, 84], deformation and water penetration into the crevices in the VSDs were observed. Interactions of the phosphate groups of the bilayer PCs with the arginine residues were also proposed in the activated state, supporting the idea that the phosphate groups favor the activated state of the VSDs. These calculations are consistent with the experimental studies of the importance of the phosphate groups to the opening of the channel and the electrostatic effects of phosphatidic acid (PA) on the VSDs from either side of the bilayer [85]. The energetic change due to the added negative charges from the PA is much weaker than the lipid-dependent gating effect from the nonphospholipids (Figure 4), probably due to the presence of phosphate groups at the boundaries of the hydrophobic core.

Solid-state NMR studies of labeled VSDs in different lipid bilayers failed to detect significant conformational changes of the VSDs between a phospholipid bilayer and a DOTAP one [86]. It could have been due to the difficulty of using local environments at the labeled sites to deduce the conformational switches of the VSDs. Lacking *ab initio* structure determination of the channel in the DOTAP bilayer or reliable models for either resting or activated states of the VSDs might also have contributed to the difficulty in obtaining good structural views from the NMR results. New structural models are needed.

7. Physiological Importance of Lipid-Dependent Gating

One could argue that the strong lipid-dependent gating effects resulted from the Group II lipids might be strongly mitigated in cell membranes where lipid composition is complicated and Group II lipids are organized into domains. On the opposite side, if the lipid-dependent gating is critical, it might be manifested in human patients carrying mutations that change lipid homeostasis. We surveyed most well-known lipid metabolic defects in humans (Table 1) [87, 88]. Surprisingly, all these defects are related to nonphospholipid homeostasis. NPA and NPB defects cause changes in sphingomyelin distribution and thus the packing of cholesterol and sphingomyelin in membranes. Our hypothesis of lipid-dependent gating predicts that altered nonphospholipid content in cell membranes would change the functional property of Kv channels whose annular lipids are affected by the altered lipid composition, and the changes in the Kv channel function may in turn lead to significant alterations in cellular excitability. It means that metabolic defects in nonphospholipids are expected to elicit changes in neuronal excitability in certain cells and may cause neurological diseases. In accord with this prediction, *all* known lipid metabolic defects in human patients bring about severe neurological defects, long before the obvious loss of white matter due to axonal demyelination.

The lipid-dependent gating effect suggests that the current paradigm for neurological diseases caused by defects in lipid metabolism might be incomplete. During the development of lipid metabolic diseases the intracellular accumulation of nonphospholipids may change the gating properties of Kv channels and alter neuronal excitability due to the altered content of nonphospholipids in cell membranes. As expected, all mutations listed in Table 1 impair the function of the central nervous system (CNS). In clinical settings, these disorders are usually diagnosed in late stages when there is clearly decreased or abnormally configured white matter in the brain as detected by MRI. However, in earlier disease stages, neurological abnormalities cannot be easily explained because of the absence of obvious demyelination. There are also other cases where altered myelination does not lead to neuronal hyperexcitability and epilepsy [89]. These two aspects suggest that demyelination explains only some aspects of the lipid metabolic defects on neurological diseases. Consequently, treatment of these diseases has been limited to manipulating specific enzymes or dietary lipids, but has generally achieved very little success such that most of these diseases lead to the early demise of afflicted patients as a result of our poor understanding. Lipid-dependent gating provides a different perspective into the neuronal defects associated with the lipid-metabolic diseases.

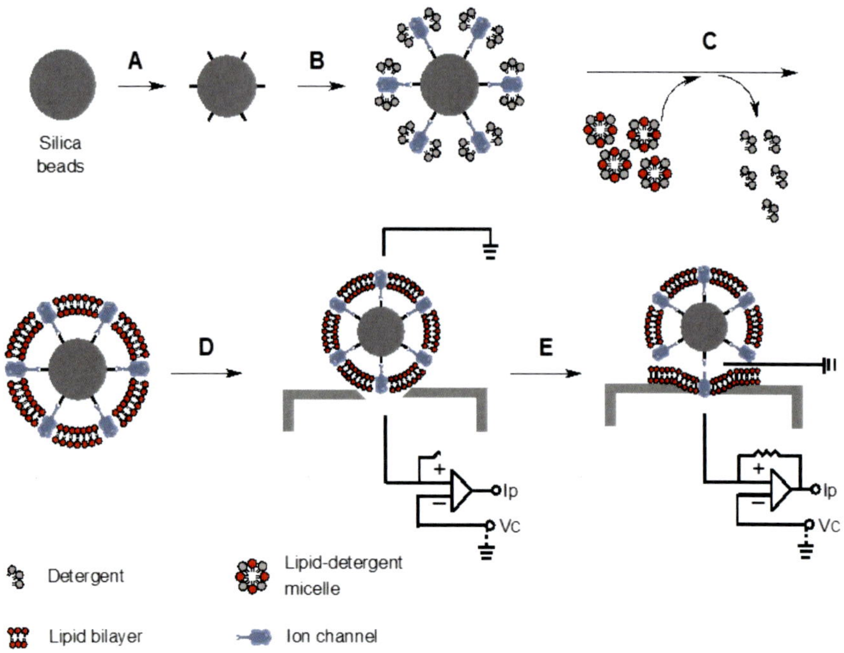

Figure 6. bSUMs as a new membrane system. Beads are chemically functionalized with a specific ligand (His_6-tag as an example; A), and then incubated with His-tagged channels (B). Detergent-solubilized lipids are added to the mixture and the detergents are slowly removed by step-wise incubation with BioBeads (C). The membrane formed around the beads is attached to a planar electrode (D). After spontaneous formation of giga-ohm seal, the membrane patch breaks due to the attachment of the membrane to the plasma-treated glass surface, and it can be used for patch-clamp recording (E).

To fully test the neuronal hypothesis in the lipid metabolic defects and the application of lipid-dependent gating to account for at least partially the altered excitability in CNS, we will need a systematic study of specific types of Kv channels in mutated cell lines and neurons from animal models carrying the specific mutations in the genome. Purkinje neurons in mouse cerebellum and dopaminergic neurons in the brain may be good systems to start these investigations. If tested successfully, lipid-dependent gating would suggest that modulators of the identified Kv channels might offer a different method to mitigate the neurological phenotypes in the affected patients.

Table 1. Lipid metabolic defects primarily cause the intracellular accumulation of non-phospholipids and severe neurological diseases in human patients

Diseases	Mutated gene	Accumulated nonphospholipids	Affected organs	Prevalence
X-linked adrenoleuko-dystrophy (X-ALD)	ABCD1 and down-regulated ALDP	Very long-chain fatty acids (VLCFAs)	Nervous systems and adrenal glands	1 in 20,000 to 50,000 worldwide
Krabbe leukodystrophy (KD)	GALC (galactosylcera-midase)	Galactocerebrosides, galactosyl-sphingosine, lactosylceramide, and monogalactosyldiglyceride	Nervous systems (CNS and PNS)	1 in 100,000 births; 6 per 1000 in some Arab communities. Often fatal.
Metachromatic leukodystrophy (MLD)	ASA (arylsulfatase)	Galactocerebroside sulfate or galactosylceramide sulfate	Mainly nervous system and kidney	1 in 40,000 to 160,000
Cerebrotendinous xanthomatosis (CTX)	CYP27A1 (sterol 27-hydroxylase)	Cholestanol	Nerves, brain, tendons	3-5 in 100,000
Fabry's disease	GLA	Glycolipids	Skin, heart, kidney, brain	1 in 40,000 to 60,000 male, less in female
Tay-Sachs disease	HEXA	GM2 gangliosides	Nervous system, muscle, etc.	1 in 90,000; mainly in infants, progressive and fatal.
Wolman's disease	LIPA (lysosomal acid lipase)	Cholesterol and glycerides	Brain, spleen, liver, adrenal gland,	1 in 500,000
Niemann-Pick disease type A, B, C	ASMD (type A, B); NPC1 and NPC2 (type C)	Sphingomyelin, cholesterol	Liver, spleen, lymph nodes, brain, etc.	Very rare, 500 cases diagnosed worldwide

8. Challenges in Improving Future Investigations of Lipid-Dependent Gating

Studies of lipid-protein interactions have benefited from multiple membrane systems where recombinant proteins can be purified and studied *in vitro* or in native membranes, and from cell-based systems where mutated proteins can be studied in cultured cells or in model animals. However, it remains difficult to quantify the lipid-protein interactions in native cell membranes because of the complicated lipid composition in eukaryotic cells, the organization of membrane proteins and lipids into dynamic domains, and the added effects from different phospholipids and nonphospholipids. Our biophysical model suggests that the annular lipids and a Kv channel may act together as a unit so that the allosteric changes in lipids and those in the channel may mutually affect each other. Currently only crystallographic studies of Ca^{2+}-ATPase and nAchR have revealed the positions for lipids in the annular layer and separated specific binding sites from nonspecific ones. Unfortunately there is no direct method that has sufficient resolutions in space and time in order to quantify dynamic lipid-channel interactions in either *in vitro* homogeneous membrane systems or in cell membranes.

In order to build a membrane system that allows easy incorporation of nonphospholipids into the bilayers, we developed the bSUM system (Figure 6) [54]. The bSUM is advantageous in six aspects: a) A majority of the ion channels is orientated to the same direction; b) Channels mediate the formation of the lipid bilayer; c) There is one and only one bilayer around each bead; d) Lipids are changeable and the bead sizes can vary from 0.1 to 20 μm; e) Channels can be recorded by patch clamp; f) Fast voltage-clamp speed (0.2-0.5 ms) is suitable for studying Kv channels with fast kinetics. In my laboratory bSUM made it possible to introduce cholesterol into phospholipid membranes at a prefixed ratio, and allowed us to record the channel activity with sufficient time resolution and quantify the lipid-induced energetic changes to the channel protein (Figure 4). Giant-patch recordings can be developed in the future in order to make recordings from ~300,000 channels and study the lipid-dependent effect on the movement of the VSDs in membranes of varying lipid composition.

The bSUM system can be made with porous beads and can be prepared with a supporting domain so that the space between the bead and the membrane will be wider. It is also suitable for super-resolution fluorescence imaging to resolve lipid distribution next to individual channel proteins [90]. Using single molecular FRET or other tools, it may also be feasible to study partition of lipids between the annular layer and the bulk phase and the dynamic interaction of the lipids with the channels at both specific and nonspecific binding sites. A point separation resolution better than 1.0 nm must be achieved in order to obtain sufficient information to differentiate nonphospholipids from phospholipids. High-resolution atomic force microscopy (AFM) at subnanometer resolution or other techniques may be further developed to satisfy such a requirement [91].

With the current development of cryoEM and tomographic studies, subnanometer tomographic study of membrane patches will likely become feasible such that the distinction between nonphospholipids and phospholipids may be sufficient to allow the resolution of all lipids around each channel (or membrane protein) [36, 92]. Alternatively, with the high-brilliance free-electron X-ray laser pulse and transient collection of diffraction data from all possible directions using direct detectors, it may become feasible to obtain sufficient data from an individual membrane patch and build *ab initio* atomic models to resolve both lipids and proteins in it [93-96]. These two techniques belong to the future and if successfully developed, will solve a lot of the fundamental questions on the lipid-dependent gating of ion channels in both *in vitro* and *in vivo* membranes.

9. General Conclusions and Future Directions

Lipid-dependent gating of Kv channels is a working hypothesis that has sustained the test of time in the past few years. Its biophysical model suggests that the annular lipids next to the channel proteins may exert strong effects and change the dynamic switching of the channels between different gating states. The annular lipids may contact with the transmembrane domain of the protein at specific or non-specific sites. The collective interaction at the nonspecific sites can together make a significant contribution to the free energy change required for shifting the equilibrium of the channels among different conformational states. More experiments are required to investigate whether the energetic difference caused by nonphospholipids vs. phospholipids is more generic to the extended family of voltage-gated ion channels and other membrane proteins that undergo major conformational changes within membranes, and whether such a difference is a key mechanism underlying the neurological defects caused by altered lipid composition in specific cells from either model animals or patients carrying lipid metabolic defects.

High-resolution tools will be needed in order to directly visualize the distribution of nonphospholipids and phospholipids in the annular layer around individual ion channels --- the key players in the lipid-depedent gating.

References

[1] Iwasa J, Marshall W. *Karp's Cell and Molecular Biology*. 8th ed. John Wiley & Sons; 2016.

[2] Hille B. *Ion Channels of Excitable Membranes*. 3rd ed. Sunderland, MA, USA: Sinauer Associates, Inc.; 2001.

[3] Gouaux E. alpha-Hemolysin from Staphylococcus aureus: an archetype of beta-barrel, channel-forming toxins. *J Struct Biol.* 1998;121 (2):110-22. PMID: 9615434.

[4] Mukherjee S, Zheng H, Derebe MG, Callenberg KM, Partch CL, Rollins D, Propheter DC, Rizo J, Grabe M, Jiang QX, Hooper LV. Antibacterial membrane attack by a pore-forming intestinal C-type lectin. *Nature*. 2013. Epub 2013/11/22. doi: 10.1038/nature12729. PMID: 24256734.

[5] Zhang S, Udho E, Wu Z, Collier RJ, Finkelstein A. Protein translocation through anthrax toxin channels formed in planar lipid bilayers. *Biophys J.* 2004;87 (6):3842-9. PMID: 15377524.

[6] Slatin SL, Qiu XQ, Jakes KS, Finkelstein A. Identification of a translocated protein segment in a voltage-dependent channel. *Nature.* 1994;371 (6493):158-61. PubMed PMID: 7521016.

[7] Sreelatha A, Bennett TL, Zheng H, Jiang QX, Orth K, Starai VJ. Vibrio effector protein, VopQ, forms a lysosomal gated channel that disrupts host ion homeostasis and autophagic flux. *Proc Natl Acad Sci US.* 2013;110 (28):11559-64. PubMed PMID: 23798441; PMCID: PMC3710849.

[8] Wang W, Whorton MR, MacKinnon R. Quantitative analysis of mammalian GIRK2 channel regulation by G proteins, the signaling lipid PIP2 and Na+ in a reconstituted system. *eLife.* 2014;3:e03671. PubMed PMID: 25049222; PMCID: PMC4135351.

[9] Zaydman MA, Silva JR, Delaloye K, Li Y, Liang H, Larsson HP, Shi J, Cui J. Kv7.1 ion channels require a lipid to couple voltage sensing to pore opening. *Proc Natl Acad Sci US.* 2013;110 (32):13180-5. PubMed PMID: 23861489; PMCID: PMC3740903.

[10] Whorton MR, MacKinnon R. X-ray structure of the mammalian GIRK2-betagamma G-protein complex. *Nature.* 2013;498 (7453): 190-7. PubMed PMID: 23739333; PMCID: PMC4654628.

[11] Lopes CM, Remon JI, Matavel A, Sui JL, Keselman I, Medei E, Shen Y, Rosenhouse-Dantsker A, Rohacs T, Logothetis DE. Protein kinase A modulates PLC-dependent regulation and PIP2-sensitivity of K+ channels. *Channels* (Austin). 2007;1 (2):124-34. PubMed PMID: 18690021.

[12] Lee J, Cha SK, Sun TJ, Huang CL. PIP2 activates TRPV5 and releases its inhibition by intracellular Mg2+. *J Gen Physiol.* 2005;126 (5):439-51. PubMed PMID: 16230466.

[13] Gamper N, Reznikov V, Yamada Y, Yang J, Shapiro MS. Phosphatidylinositol [correction] 4,5-bisphosphate signals underlie receptor-specific Gq/11-mediated modulation of N-type Ca2+ channels. *J Neurosci.* 2004;24 (48):10980-92. PubMed PMID: 15574748.

[14] Xiao J, Zhen XG, Yang J. Localization of PIP2 activation gate in inward rectifier K+ channels. *Nat Neurosci.* 2003;6 (8):811-8. PubMed PMID: 12858177.

[15] Bian J, Cui J, McDonald TV. HERG K (+) channel activity is regulated by changes in phosphatidyl inositol 4,5-bisphosphate. *Circ Res.* 2001;89 (12):1168-76. PubMed PMID: 11739282.

[16] Valiyaveetil FI, Zhou Y, MacKinnon R. Lipids in the structure, folding, and function of the KcsA K+ channel. *Biochemistry.* 2002;41 (35):10771-7 3149. PubMed PMID: 3330.

[17] Powl AM, East JM, Lee AG. Anionic phospholipids affect the rate and extent of flux through the mechanosensitive channel of large conductance MscL. *Biochemistry.* 2008; 47 (14):4317-28. PubMed PMID: 18341289.

[18] Boland LM, Drzewiecki MM. Polyunsaturated fatty acid modulation of voltage-gated ion channels. *Cell Biochem Biophys*. 2008;52 (2):59-84. PubMed PMID: 18830821.

[19] Yang F, Zheng J. Understand spiciness: mechanism of TRPV1 channel activation by capsaicin. *Protein Cell*. 2017;8 (3):169-77. PubMed PMID: 28044278; PMCID: PMC5326624.

[20] Taylor KC, Sanders CR. Regulation of KCNQ/Kv7 family voltage-gated K+ channels by lipids. *Biochim Biophys Acta*. 2017;1859 (4):586-97. PubMed PMID: 27818172; PMCID: PMC5305565.

[21] Elinder F, Liin SI. Actions and Mechanisms of Polyunsaturated Fatty Acids on Voltage-Gated Ion Channels. *Front Physiol*. 2017;8:43. PubMed PMID: 28220076; PMCID: PMC5292575.

[22] Cuppoletti J, Tewari KP, Chakrabarti J, Malinowska DH. Identification of the fatty acid activation site on human ClC-2. *Am J Physiol Cell Physiol*. 2017;312 (6):C707-C23. PubMed PMID: 28424169.

[23] Dimitrova A, Walko M, Hashemi Shabestari M, Kumar P, Huber M, Kocer A. In situ, Reversible Gating of a Mechanosensitive Ion Channel through Protein-Lipid Interactions. *Frontiers in physiology*. 2016;7:409. PubMed PMID: 27708587; PMCID: PMC5030285.

[24] Ben-Zeev G, Telias M, Nussinovitch I. Lysophospholipids modulate voltage-gated calcium channel currents in pituitary cells; effects of lipid stress. *Cell Calcium*. 2010;47 (6):514-24. PubMed PMID: 20510448.

[25] Schilling T, Eder C. Non-selective cation channel activity is required for lysophosphatidylcholine-induced monocyte migration. *J Cell Physiol*. 2009;221 (2):325-34. PubMed PMID: 19562685.

[26] Burnashev NA, Undrovinas AI, Fleidervish IA, Makielski JC, Rosenshtraukh LV. Modulation of cardiac sodium channel gating by lysophosphatidylcholine. *J Mol Cell Cardiol*. 1991;23 Suppl 1:23-30. PubMed PMID: 1645412.

[27] Pristera A, Baker MD, Okuse K. Association between tetrodotoxin resistant channels and lipid rafts regulates sensory neuron

excitability. *PLoS ONE*. 2012;7 (8):e40079. PubMed PMID: 2287 0192; PMCID: PMC3411591.

[28] Lee AK, Yeung-Yam-Wah V, Tse FW, Tse A. Cholesterol elevation impairs glucose-stimulated Ca (2+) signaling in mouse pancreatic beta-cells. *Endocrinology*. 2011;152 (9):3351-61. PubMed PMID: 21712366.

[29] Levitan I, Fang Y, Rosenhouse-Dantsker A, Romanenko V. Cholesterol and ion channels. *Subcell Biochem*. 2010;51:509-49.. PubMed PMID: 20213557; PMCID: 2895485.

[30] Balijepalli RC, Delisle BP, Balijepalli SY, Foell JD, Slind JK, Kamp TJ, January CT. Kv11.1 (ERG1) K+ channels localize in cholesterol and sphingolipid enriched membranes and are modulated by membrane cholesterol. *Channels* (Austin). 2007;1 (4):263-72. PubMed PMID: 18708743.

[31] Davies A, Douglas L, Hendrich J, Wratten J, Tran Van Minh A, Foucault I, Koch D, Pratt WS, Saibil HR, Dolphin AC. The calcium channel alpha2delta-2 subunit partitions with CaV2.1 into lipid rafts in cerebellum: implications for localization and function. *J Neurosci*. 2006;26 (34):8748-57. PubMed PMID: 16928863.

[32] Xia F, Gao X, Kwan E, Lam PP, Chan L, Sy K, Sheu L, Wheeler MB, Gaisano HY, Tsushima RG. Disruption of pancreatic beta-cell lipid rafts modifies Kv2.1 channel gating and insulin exocytosis. *J Biol Chem*. 2004;279 (23):24685-91. PubMed PMID: 15073181.

[33] Martens JR, O'Connell K, Tamkun M. Targeting of ion channels to membrane microdomains: localization of KV channels to lipid rafts. *Trends Pharmacol Sci*. 2004;25 (1):16-21. PubMed PMID: 147 23974.

[34] Hajdu P, Varga Z, Pieri C, Panyi G, Gaspar R, Jr. Cholesterol modifies the gating of Kv1.3 in human T lymphocytes. *Pflugers Arch*. 2003;445 (6):674-82. PubMed PMID: 12632187.

[35] Ruprecht JJ, Mielke T, Vogel R, Villa C, Schertler GF. Electron crystallography reveals the structure of metarhodopsin I. *EMBO J*. 2004;23 (18):3609-20 2235. PubMed PMID: 2761.

[36] Unwin N. Segregation of lipids near acetylcholine-receptor channels imaged by cryo-EM. *IUCrJ.* 2017;4 (Pt 4):393-9. PubMed PMID: 28875026; PMCID: PMC5571802.

[37] Sun J, Comeau JF, Baenziger JE. Probing the structure of the uncoupled nicotinic acetylcholine receptor. *Biochim Biophys Acta.* 2017;1859 (2):146-54.. PubMed PMID: 27871840.

[38] Baenziger JE, Domville JA, Therien JPD. The Role of Cholesterol in the Activation of Nicotinic Acetylcholine Receptors. *Curr Top Membr*. 2017;80:95-137. doi: 10.1016/bs.ctm.2017.05.002. PubMed PMID: 28863823.

[39] Fantini J, Barrantes FJ. Sphingolipid/cholesterol regulation of neurotransmitter receptor conformation and function. *Biochim Biophys Acta*. 2009;1788 (11):2345-61. PubMed PMID: 19733149.

[40] Brannigan G, Henin J, Law R, Eckenhoff R, Klein ML. Embedded cholesterol in the nicotinic acetylcholine receptor. *Proc Natl Acad Sci US.* 2008;105 (38):14418-23. PubMed PMID: 18768796; PMCID: PMC2567199.

[41] Barrantes FJ. Cholesterol effects on nicotinic acetylcholine receptor. *J Neurochem*. 2007;103 Suppl 1:72-80. 2007.04719.x. PubMed PMID: 17986142.

[42] Stetzkowski-Marden F, Recouvreur M, Camus G, Cartaud A, Marchand S, Cartaud J. Rafts are required for acetylcholine receptor clustering. *Journal of molecular neuroscience*: MN. 2006;30 (1-2):37-8. PubMed PMID: 17192619.

[43] Barrantes FJ. Lipid matters: nicotinic acetylcholine receptor-lipid interactions (Review). *Mol Membr Biol.* 2002;19 (4):277-84. PubMed PMID: 12512774.

[44] Brannigan G. Direct Interactions of Cholesterol With Pentameric Ligand-Gated Ion Channels: Testable Hypotheses From Computational Predictions. *Curr Top Membr*. 2017;80:163-86. PubMed PMID: 28863815.

[45] Henin J, Salari R, Murlidaran S, Brannigan G. A predicted binding site for cholesterol on the GABAA receptor. *Biophys J.* 2014;106 (9):1938-49. PubMed PMID: 24806926; PMCID: PMC4017285.

[46] Rosenhouse-Dantsker A, Logothetis DE, Levitan I. Cholesterol sensitivity of KIR2.1 is controlled by a belt of residues around the cytosolic pore. *Biophys J.* 2011;100 (2):381-9. PubMed PMID: 21244834; PMCID: PMC3021658.

[47] Levitan I. Cholesterol and Kir channels. *IUBMB life*. 2009;61 (8):781-90. PubMed PMID: 19548316; PMCID: PMC2720429.

[48] Levitan I, Singh DK, Rosenhouse-Dantsker A. Cholesterol binding to ion channels. *Frontiers in physiology*. 2014;5:65. PubMed PMID: 24616704; PMCID: PMC3935357.

[49] Dopico AM, Bukiya AN. Lipid regulation of BK channel function. *Frontiers in physiology*. 2014;5:312. PubMed PMID: 25202277; PMCID: 4141547.

[50] Zaydman MA, Cui J. PIP2 regulation of KCNQ channels: biophysical and molecular mechanisms for lipid modulation of voltage-dependent gating. *Front Physiol.* 2014;5:195. PubMed PMID: 24904429; PMCID: PMC4034418.

[51] Zheng H, Liu W, Anderson LY, Jiang QX. Lipid-dependent gating of a voltage-gated potassium channel. *Nat Commun*. 2011;2:250. Epub 2011/03/24. PubMed PMID: 21427721; PMCID: 3072105.

[52] Xia F, Xie L, Mihic A, Gao X, Chen Y, Gaisano HY, Tsushima RG. Inhibition of cholesterol biosynthesis impairs insulin secretion and voltage-gated calcium channel function in pancreatic beta-cells. *Endocrinology.* 2008;149 (10):5136-45. PubMed PMID: 18599549.

[53] Norimatsu Y, Hasegawa K, Shimizu N, Toyoshima C. Protein-phospholipid interplay revealed with crystals of a calcium pump. *Nature.* 2017;545 (7653):193-8. PubMed PMID: 28467821.

[54] Zheng H, Lee S, Llaguno MC, Jiang QX. bSUM: A bead-supported unilamellar membrane system facilitating unidirectional insertion of membrane proteins into giant vesicles. *J Gen Physiol*. 2016;147 (1):77-93. PubMed PMID: 26712851; PMCID: PMC4692488.

[55] Jiang QX, Gonen T. The influence of lipids on voltage-gated ion channels. *Curr Opin Struct Biol*. 2012;22 (4):529-36. Epub 2012/04/10. PubMed PMID: 22483432; PMCID: PMC3408884.

[56] Schmidt D, Jiang QX, MacKinnon R. Phospholipids and the origin of cationic gating charges in voltage sensors. *Nature.* 2006;444 (7120):775-9. PubMed PMID: 17136096.

[57] Jiang Q-X, Gonen T. The influence of lipids on voltage-gated ion channels. *Curr Opin Struct Biol.* 2012;22:529-36. PubMed PMID: 22483432; PMCID: PMC3408884.

[58] Xu Y, Ramu Y, Lu Z. Removal of phospho-head groups of membrane lipids immobilizes voltage sensors of K+ channels. *Nature*. 2008;451 (7180):826-9. PubMed PMID: 18273018.

[59] O'Connell KM, Loftus R, Tamkun MM. Localization-dependent activity of the Kv2.1 delayed-rectifier K+ channel. *Proc Natl Acad Sci US.* 2010;107 (27):12351-6. PubMed PMID: 20566856; PMCID: 2901471.

[60] Randich AM, Cuello LG, Wanderling SS, Perozo E. Biochemical and structural analysis of the hyperpolarization-activated K (+) channel MVP. *Biochemistry.* 2014;53 (10):1627-36. PubMed PMID: 24490868; PMCID: 3985891.

[61] Purcell EK, Liu L, Thomas PV, Duncan RK. Cholesterol influences voltage-gated calcium channels and BK-type potassium channels in auditory hair cells. *PLoS ONE.* 2011;6 (10):e26289. Epub 2011/11/03. PubMed PMID: 22046269; PMCID: 3194812.

[62] Huang CW, Wu YJ, Wu SN. Modification of activation kinetics of delayed rectifier K+ currents and neuronal excitability by methyl-beta-cyclodextrin. *Neuroscience.* 2011;176:431-41. PubMed PMID: 21073928.

[63] Finol-Urdaneta RK, McArthur JR, Juranka PF, French RJ, Morris CE. Modulation of KvAP unitary conductance and gating by 1-alkanols and other surface active agents. *Biophys J.* 2010;98 (5):762-72. PubMed PMID: 20197029; PMCID: 2830433.

[64] Guo J, Chi S, Xu H, Jin G, Qi Z. Effects of cholesterol levels on the excitability of rat hippocampal neurons. *Mol Membr Biol*. 2008;25 (3):216-23. PubMed PMID: 18428037.

[65] Pottosin, II, Valencia-Cruz G, Bonales-Alatorre E, Shabala SN, Dobrovinskaya OR. Methyl-beta-cyclodextrin reversibly alters the

gating of lipid rafts-associated Kv1.3 channels in Jurkat T lymphocytes. *Pflugers Arch.* 2007;454 (2):235-44. PubMed PMID: 17242956.

[66] Abi-Char J, Maguy A, Coulombe A, Balse E, Ratajczak P, Samuel JL, Nattel S, Hatem SN. Membrane cholesterol modulates Kv1.5 potassium channel distribution and function in rat cardiomyocytes. *J Physiol.* 2007;582 (Pt 3):1205-17. PubMed PMID: 17525113; PMCID: PMC2075263.

[67] Pouvreau S, Berthier C, Blaineau S, Amsellem J, Coronado R, Strube C. Membrane cholesterol modulates dihydropyridine receptor function in mice fetal skeletal muscle cells. *J Physiol.* 2004;555 (Pt 2):365-81. PubMed PMID: 14724204.

[68] Rudakova E, Wagner M, Frank M, Volk T. Localization of Kv4.2 and KChIP2 in lipid rafts and modulation of outward K+ currents by membrane cholesterol content in rat left ventricular myocytes. *Pflugers Arch.* 2015;467 (2):299-309. PubMed PMID: 24793047.

[69] Bowles DK, Heaps CL, Turk JR, Maddali KK, Price EM. Hypercholesterolemia inhibits L-type calcium current in coronary macro-, not microcirculation. *J Appl Physiol* (1985). 2004;96 (6):2240-8. PubMed PMID: 14752123.

[70] Heaps CL, Tharp DL, Bowles DK. Hypercholesterolemia abolishes voltage-dependent K+ channel contribution to adenosine-mediated relaxation in porcine coronary arterioles. *Am J Physiol Heart Circ Physiol.* 2005;288 (2):H568-76. PubMed PMID: 15458946.

[71] Balajthy A, Somodi S, Petho Z, Peter M, Varga Z, Szabo GP, Paragh G, Vigh L, Panyi G, Hajdu P. 7DHC-induced changes of Kv1.3 operation contributes to modified T cell function in Smith-Lemli-Opitz syndrome. *Pflugers Arch.* 2016 PubMed PMID: 27315086.

[72] Chun YS, Shin S, Kim Y, Cho H, Park MK, Kim TW, Voronov SV, Di Paolo G, Suh BC, Chung S. Cholesterol modulates ion channels via down-regulation of phosphatidylinositol 4,5-bisphosphate. *J Neurochem.* 2010;112 (5):1286-94. PubMed PMID: 20015154; PMCID: 2891522.

[73] Coyan FC, Abderemane-Ali F, Amarouch MY, Piron J, Mordel J, Nicolas CS, Steenman M, Merot J, Marionneau C, Thomas A, Brasseur R, Baro I, Loussouarn G. A long QT mutation substitutes cholesterol for phosphatidylinositol-4,5-bisphosphate in KCNQ1 channel regulation. *PLoS ONE*. 2014;9 (3):e93255. PubMed PMID: 24681627; PMCID: PMC3969324.
[74] Calhoun JD, Isom LL. The role of non-pore-forming beta subunits in physiology and pathophysiology of voltage-gated sodium channels. *Handb Exp Pharmacol*. 2014;221:51-89. PubMed PMID: 24737232.
[75] Catterall WA. Molecular properties of voltage-sensitive sodium and calcium channels. *Braz J Med Biol Res*. 1988;21 (6):1129-44. PubMed PMID: 2855029.
[76] Hofmann F, Belkacemi A, Flockerzi V. Emerging Alternative Functions for the Auxiliary Subunits of the Voltage-Gated Calcium Channels. *Curr Mol Pharmacol*. 2015;8 (2):162-8. PubMed PMID: 25966689.
[77] Wu J, Yan Z, Li Z, Yan C, Lu S, Dong M, Yan N. Structure of the voltage-gated calcium channel Cav1.1 complex. *Science*. 2015;350 (6267):aad2395. PubMed PMID: 26680202.
[78] Hite RK, Li Z, Walz T. Principles of membrane protein interactions with annular lipids deduced from aquaporin-0 2D crystals. *Embo J*. 2010;29 (10):1652-8.
[79] Gonen T, Sliz P, Kistler J, Cheng Y, Walz T. Aquaporin-0 membrane junctions reveal the structure of a closed water pore. *Nature*. 2004;429 (6988):193-7 844. PubMed PMID: 2802.
[80] Long SB, Tao X, Campbell EB, MacKinnon R. Atomic structure of a voltage-dependent K+ channel in a lipid membrane-like environment. *Nature*. 2007;450 (7168):376-82. PubMed PMID: 18004376.
[81] Nyholm TK, Lindroos D, Westerlund B, Slotte JP. Construction of a DOPC/PSM/Cholesterol Phase Diagram Based on the Fluorescence Properties of trans-Parinaric Acid. *Langmuir*. 2011. Epub 2011/06/02. PubMed PMID: 21627141.
[82] Andersson M, Freites JA, Tobias DJ, White SH. Structural dynamics of the S4 voltage-sensor helix in lipid bilayers lacking phosphate

groups. *J Phys Chem B.* 2011;115 (27):8732-8. PubMed PMID: 21692541; PMCID: PMC3140535.

[83] Sands ZA, Sansom MS. How does a voltage sensor interact with a lipid bilayer? Simulations of a potassium channel domain. *Structure.* 2007;15 (2):235-44. PubMed PMID: 17292841; PMCID: PMC18 85962.

[84] Bond PJ, Sansom MS. Bilayer deformation by the Kv channel voltage sensor domain revealed by self-assembly simulations. *Proc Natl Acad Sci US.* 2007;104 (8):2631-6. PubMed PMID: 17301243; PMCID: PMC1797625.

[85] Hite RK, Butterwick JA, MacKinnon R. Phosphatidic acid modulation of Kv channel voltage sensor function. *eLife.* 2014;3. PubMed PMID: 25285449.

[86] Krepkiy D, Gawrisch K, Swartz KJ. Structural interactions between lipids, water and S1-S4 voltage-sensing domains. *J Mol Biol.* 2012;423 (4):632-47. PubMed PMID: 22858867; PMCID: PMC3616881.

[87] Jia Z, Pei Z, Li Y, Wei L, Smith KD, Watkins PA. X-linked adrenoleukodystrophy: role of very long-chain acyl-CoA synthetases. *Mol Genet Metab.* 2004;83 (1-2):117-27. PubMed PMID: 15464426.

[88] Lu JF, Lawler AM, Watkins PA, Powers JM, Moser AB, Moser HW, Smith KD. A mouse model for X-linked adrenoleukodystrophy. *Proc Natl Acad Sci US.* 1997;94 (17):9366-71. PubMed PMID: 9256488; PMCID: PMC23196.

[89] Pascual JM, Campistol J, Gil-Nagel A. Epilepsy in inherited metabolic disorders. *Neurologist.* 2008;14 (6 Suppl 1):S2-S14. PubMed PMID: 19225367.

[90] Balzarotti F, Eilers Y, Gwosch KC, Gynna AH, Westphal V, Stefani FD, Elf J, Hell SW. Nanometer resolution imaging and tracking of fluorescent molecules with minimal photon fluxes. *Science.* 2017;355 (6325):606-12. PubMed PMID: 28008086.

[91] Pfreundschuh M, Harder D, Ucurum Z, Fotiadis D, Muller DJ. Detecting Ligand-Binding Events and Free Energy Landscape while

Imaging Membrane Receptors at Subnanometer Resolution. *Nano letters.* 2017;17 (5):3261-9. PubMed PMID: 28361535.

[92] Miao J, Ercius P, Billinge SJ. Atomic electron tomography: 3D structures without crystals. *Science.* 2016;353 (6306). PubMed PMID: 27708010.

[93] Song C, Jiang H, Mancuso A, Amirbekian B, Peng L, Sun R, Shah SS, Zhou ZH, Ishikawa T, Miao J. Quantitative imaging of single, unstained viruses with coherent x rays. *Phys Rev Lett.* 2008;101 (15):158101. PubMed PMID: 18999646.

[94] Miao J, Hodgson KO, Ishikawa T, Larabell CA, LeGros MA, Nishino Y. Imaging whole Escherichia coli bacteria by using single-particle x-ray diffraction. *Proc Natl Acad Sci US.* 2003;100 (1):110-2. Epub 2003/01/09. PubMed PMID: 12518059; PMCID: 140897.

[95] Leshem B, Xu R, Dallal Y, Miao J, Nadler B, Oron D, Dudovich N, Raz O. Direct single-shot phase retrieval from the diffraction pattern of separated objects. *Nat Commun.* 2016;7:10820. PubMed PMID: 26899582; PMCID: PMC4764927.

[96] Schlichting I, Miao J. Emerging opportunities in structural biology with X-ray free-electron lasers. *Curr Opin Struct Biol.* 2012;22 (5):613-26. PubMed PMID: 22922042; PMCID: PMC3495068.

In: Protein-Lipid Interactions
Editor: Angel Catalá
ISBN: 978-1-53613-125-3

Chapter 5

USING NOVEL IMAGING APPROACHES TO STUDY PHOSPHOINOSITIDE SIGNALING IN VESICLE TRAFFICKING

*Xuelin Lou**, *MD PhD*
Department of Neuroscience, University of Wisconsin–Madison, School of Medicine and Public Health, Madison, WI, US

ABSTRACT

Phosphoinositides (PIs) contribute only a small portion of membrane lipids but are essential for cell function. Dysfunction of PIs metabolism and signaling is increasingly implicated in a number of human diseases. Seven types of PIs equilibrate dynamically in living cells and have compartmentalized distribution in various cellular organelles. Such spatial organization pattern facilitates local protein interactions for organelle-specific function in time and space. As a second messenger and the precursor of DAG-PKC/Munc13 signaling pathway, $PI(4,5)P_2$ locates mainly in the inner leaflet of the plasma membrane and participates in diverse cellular functions. Here we focus on the PI signaling in vesicle trafficking of secretory cells and summarize the molecular mechanisms by which PIs regulate vesicle mobilization, docking, priming, fusion, clathrin-coat formation, endocytosis, and vesicle uncoating. Moreover, we discuss recent cutting-edge approaches that allow investigating PI

* Corresponding Author Email: xlou188@gmail.com.

signaling in a temporal-spatially defined manner during membrane trafficking. The novel technique development compensates traditional methods in biochemistry, pharmacology, and genetics, opening a new avenue to the research of lipid signaling and regulation at physiological conditions.

Keywords: phosphoinositides, $PI(4,5)P_2$, exocytosis, endocytosis, vesicle trafficking, single-molecule localization microscopy, super-resolution optical imaging, TIRF, optogenetics, cell signaling

Phosphoinositide Metabolism and Sub-Cellular Distribution

Cellular membrane is composed of two major components: protein and lipid. Phosphoinositides (PIs) are phosphorylated derivatives of phosphatidylinositol and constitute only a small fraction of cellular phospholipids. PIs are generated from reversible phosphorylation of the phosphatidylinositol ring at positions 3, 4, 5, and in combinations. There are 7 different PI species, including phosphatidylinositol 3-phosphate (PI3P), phosphatidylinositol 4-phosphate (PI4P), phosphatidylinositol 5-phosphate (PI5P), phosphatidylinositol 4,5-bisphosphate ($PI(4,5)P_2$), phosphatidylinositol 3,5-bisphosphate ($PI(3,5)P_2$), phosphatidylinositol 3,4-bisphosphate ($PI(3,4)P_2$), and phosphatidylinositol 3,4,5-trisphosphate ($PI(3,4,5)P_3$) (Figure 1A). A large number of PI kinases and phosphatases participate in the metabolism network of PIs (Balla, 2013; Di Paolo and De Camilli, 2006; Vicinanza et al., 2008) and controls the relative abundance and dynamic equilibrium of seven PIs.

All seven PIs are "active" in their own right, in addition to serving as metabolic intermediates in cells. Each of them localizes in distinct sub-cellular membrane compartments (Figure 1B). For instance, phosphatidylinositol is enriched in endoplasmic reticulum (ER), mono-phosphorylated inositides in internal membranes, and $PI(4,5)P_2$ in the plasma membrane (PM) (Balla et al., 2009). The compartmentalized distribution pattern of PIs, together with their high turnover rates, makes them suitable to act as signaling molecules in diverse cellular processes. In addition, this also provides a lipid base for organelle identity.

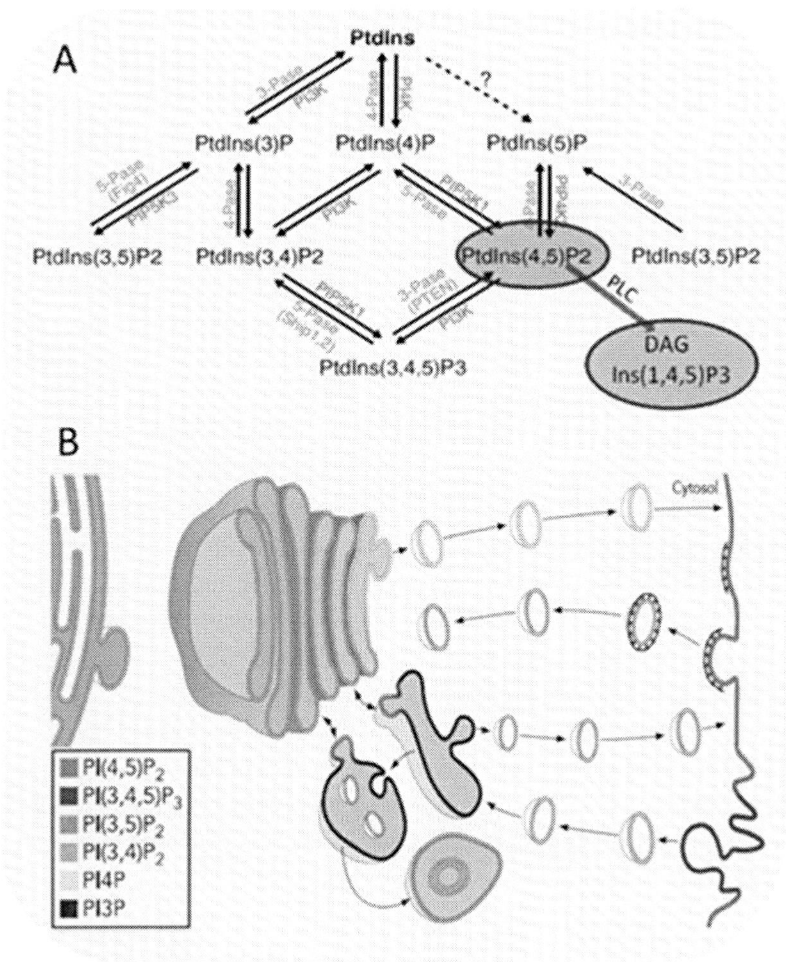

Figure 1. Metabolic network of PIs and their sub-cellular locations. (A) PI metabolic network and catalytic enzymes in cells. (B) Sub-cellular location of PIs during membrane trafficking as indicated by different grey scales. (Adapted from Vicinanza et al., and McCrea, et al.,).

PHOSPHOINOSITIDE SIGNALING AND ITS IMPLICATION IN DISEASES

PIs are increasingly recognized to play a central role in cell regulation. They interact directly and indirectly with their effectors at right time and location. It is clear now that PIs regulate broad cellular functions, such as ion channel and transporter activity (Balla, 2013; Suh and Hille, 2008b), actin cytoskeleton dynamics (Yin and Janmey, 2003), intracellular signaling (e.g., second messengers and PI3 kinase cascades) (Balla, 2013), membrane trafficking (e.g., endosome/lysosome/autophagic trafficking, exocytosis, and endocytosis) (De Camilli et al., 1996; Di Paolo and De Camilli, 2006), and nuclear signaling (Barlow et al., 2010; Viiri et al., 2012). Given the important role of PIs in cells, it is not surprising that aberrant PI metabolism and signaling have been implicated in a broad spectrum of human diseases (Balla, 2013; Halstead et al., 2005; Majerus and York, 2009; McCrea and De Camilli, 2009).

PIs are implicated in many brain diseases and metabolic disorders. Several genes encoding PI metabolic enzymes, including PIK3C2 (alpha PI3-kinase), PIK4Cα (PI4-kinase type III alpha/Stt4), PIP5K2α (a PI4P 5-kinase) and synaptojanin-1, are located in the chromosomal regions where schizophrenia, bipolar disorder or Down syndrome are mapped (McCrea and De Camilli, 2009). Accordingly, Down syndrome patients display a higher level synaptojanin-1 expression and a lower PI(4,5)P_2 level in their cerebral cortex (Arai et al., 2002) because of chromosome 21 (which harbors synaptojanin1 gene) triplication. The latter may link to early onset Alzheimer's disease (AD) in Down syndrome patients (Voronov et al., 2008). In addition, Aβ-42 peptide stimulates PI(4,5)P_2 cleavage and inhibits hippocampal long-term potentiation (LTP), and these effects can be reversed in synaptojanin1 heterozygous mice (Berman et al., 2008; Landman et al., 2006). These data highlight a potential link between aberrant PI(4,5)P_2 signaling and AD (Berman et al., 2008; Di Paolo and Kim, 2011).

In addition, strong seizure-like behaviors have been observed in mice lacking synaptojanin-1 (Cremona et al., 1999) or the synaptojanin-1 binding partner endophilin (Milosevic et al., 2011), indicating its possible involvement in network homeostasis and epilepsy *in vivo*. Genetic perturbation of PI4P 5-Kinase alpha (PI4P5K-α) significantly in beta cells alters insulin secretion and prevents obesity in mice model (Huang et al., 2011).

While the mechanistic connections between abnormal PI metabolism and phenotypic manifestations of diseases are under active investigation, drugs targeting phospholipid metabolism are already developed (Di Paolo and Kim, 2011). Deep understanding of PI signaling at the molecule, cellular, and system levels is clearly required for both basic research and clinical therapies.

PHOSPHOINOSITIDES REGULATE EXOCYTOSIS

It is well known that $PI(4,5)P_2$ is a substrate of the classical second messenger pathway. Phospholipase C (PLC) catalyzes $PI(4,5)P_2$ hydrolysis to produce soluble inositol 1,4,5-trisphosphate (IP_3) and diacylglycerol (DAG). DAG modulates exocytosis by direct interactions with the proteins containing C_1- and C_2-domains, including protein kinase C and D families (PKCs and PKDs), Munc-13, and DAG kinase (Brose and Rosenmund, 2002; Wojcik and Brose, 2007). Direct Ca^{2+} uncaging at central synapses demonstrate that phorbol esters, which mimics diacylglycerol (DAG) action, potentiates neurotransmission by directly elevating Ca^{2+} sensitivity for vesicle fusion (Lou et al., 2005). This work reveals a generic mechanism for exocytosis regulation. Downstream DAG, both PKC and Munc-13 signaling pathways contribute to the Ca^{2+} sensitivity increase (Lou et al., 2008; Wierda et al., 2007). Moreover, $PI(4,5)P_2$ itself is a signaling molecule that participates directly in multiple steps of vesicle trafficking (Ji et al., 2017).

During the regulated exocytosis, two major types of vesicles have been studied extensively. One is small clear synaptic vesicles (SVs), which are essential for synaptic transmission at nerve terminals, and the other is dense core vesicles (DCVs), which contain hormones or/and neuropeptides in secretory cells. The role of $PI(4,5)P_2$ in SV trafficking is supported by both biochemical and genetic studies. PI(4)P 5-kinase type 1γ (PIPK1γ) and synaptojanin-1, two key enzymes of $PI(4,5)P_2$ metabolism, are concentrated in central nerve terminals (Cremona et al., 1999; Wenk et al., 2001). Their gene mutations lead to defects of synaptic function and aberrant behaviors in mice (Cremona et al., 1999; Di Paolo et al., 2004; Harris et al., 2000; Mani et al., 2007). Accordingly, studies *in-vitro* and in endocrine cells support the essential role of $PI(4,5)P_2$ in exocytosis.

Because of the large size and slow dynamics of DCVs, $PI(4,5)P_2$ regulation has been extensively studied in the DCV trafficking. This line of research supports the interactions of $PI(4,5)P_2$ with multiple proteins during vesicle trafficking (Martin, 2015). For example, $PI(4,5)P_2$ regulates DCV priming in endocrine cells (Aikawa and Martin, 2003; Gong et al., 2005; Holz et al., 2000; James et al., 2008; Lawrence and Birnbaum, 2003; Milosevic et al., 2005; Olsen et al., 2003). This function relies on $PI(4,5)P_2$ interactions with CAPS1 (Grishanin et al., 2004; James et al., 2008; Loyet et al., 1998; Martin, 2001, 2015), secretory carrier membrane protein 2 (SCAMP-2) (Liao et al., 2007) or actin (Berberian et al., 2009; Wen et al., 2011).

Furthermore, $PI(4,5)P_2$ modulates membrane fusion kinetics, likely via interactions with soluble N-ethylmaleimide-sensitive factor attachment protein receptors (SNAREs) (Han et al., 2004; Jahn and Scheller, 2006) and synaptotagmin-1 (Chapman, 2008; Han et al., 2004; Pang et al., 2010; Saraswati et al., 2007; Yoshihara et al., 2010). $PI(4,5)P_2$ binds the C_2B domain of synaptotagmin in a Ca^{2+}-independent manner *in vitro* so that synaptotagmin can rapidly penetrate into the membrane (Bai et al., 2004) and accelerate liposome fusion (Arac et al., 2006; Bai et al., 2004; Stein et al., 2007). Syntaxin-1 (Rickman et al., 2010; Sieber et al., 2007) is reported to co-localized with $PI(4,5)P_2$-enriched domains (van den Bogaart et al., 2011), implicating direct interactions between $PI(4,5)P_2$ and fusion machinery syntaxin-1 during exocytosis (Aoyagi et al., 2005; Murray and Tamm, 2009). Accordingly, disruption of $PI(4,5)P_2$ binding to syntaxin-1A (Lam et al., 2008) and VAMP-2 (Williams et al., 2009) abolishes or inhibits DCV exocytosis in PC-12 cells and insulin-secreting cells, respectively.

$PI(4,5)P_2$ can also regulates exocytosis through Ca^{2+} channel modulation (Suh and Hille, 2008a; Suh et al., 2010). This modulation is indirect but very efficient because of the supra-linear Ca^{2+} dependence of vesicle fusion at central synapses (for SVs) (Lou et al., 2005; Neher and Sakaba, 2008) and in endocrine cells (for DCVs) (Voets, 2000; Voets et al., 2001).

As discussed above, $PI(4,5)P_2$ represents only a small part of PI metabolic network (Figure 1). The involvement of other PIs (and their metabolites) in vesicle trafficking remains largely unexplored, and their roles are emerging from recent studies. For example, PI3P regulates DCV secretion (Dominguez et al., 2011; Meunier et al., 2005; Osborne et al.,

2008) and GLUT4 containing vesicle insertion into the PM (Falasca et al., 2007). PI4P, Inositol hexakisphosphate (IP_6), and its pyrophosphate derivatives (IP_7) have been implicated in insulin granule secretion (Bhandari et al., 2008; Hoy et al., 2003; Illies et al., 2007; Olsen et al., 2003; Waselle et al., 2005). Future work illuminating the specific roles of PIs at vesicle trafficking will reveal new insights into cell biology and pathophysiology of neurons and endocrine cells (Frere et al., 2012).

PHOSPHOINOSITIDES REGULATE ENDOCYTOSIS

Endocytosis is a key step of vesicle trafficking in cells. PI(4,5)P_2 participates in membrane and receptor endocytosis (Posor et al., 2015), as demonstrated in two types of work. First, PI(4,5)P_2 binding motifs, such as the PH domain of PLCδ1 ($PH_{PLC\delta 1}$) and the ANTH/ENTH domain, are present in many endocytosis proteins including AP2 (Gaidarov and Keen, 1999), AP180/CALM(Ford et al., 2001; Itoh et al., 2001), amphiphysin, epsin and dynamin (Czech, 2003; Di Paolo and De Camilli, 2006). Second, perturbations of PI(4,5)P_2 metabolic enzymes impair endocytosis (Abe et al., 2008; Cremona et al., 1999; Di Paolo et al., 2004; Mani et al., 2007; Schuske et al., 2003; Zoncu et al., 2007). PI(4,5)P_2 is thought as a coincidence detector of endocytosis by recruiting endocytosis proteins (Di Paolo and De Camilli, 2006).

Moreover, PI(4,5)P_2 may directly regulate membrane fission at the neck of clathrin-coated pits, where high levels of PI(4,5)P_2 can be concentrated by dynamin PH domain (Bethoney et al., 2009; Chang-Ileto et al., 2011) or via phosphorylation-dependent AP2 binding (Nakano-Kobayashi et al., 2007). Fluorescent imaging in live cells has demonstrated the essential role of PI(4,5)P_2 metabolism mediated by synaptojanin1 in clathrin-mediated endocytosis (Abe et al., 2008; Mani et al., 2007; Perera et al., 2006). Synaptojanin1 is recruited to endocytosis sites via endophilin and disassociated from new vesicles after their uncoating. Genetic disruption of endophilin severely impairs SV uncoating, mimicking the phenotypes of synaptojanin-1 knockout (KO) synapses. These data suggest a critical role of PI(4,5)P_2 metabolism at the step of vesicle uncoating in synaptic vesicle cycle.

Using Optogenetic Approaches to Dissect Phosphoinositide Signaling in Vesicle Trafficking

During transmitter release and hormone secretion, secretory vesicles undergo several trafficking steps prior to release their cargos: vesicle mobilization toward the PM from a reserve vesicle pool, tethering/docking to the PM, molecular and positional priming, and vesicle fusion upon Ca^{2+} elevation (Imig et al., 2014; Neher and Sakaba, 2008; Rettig and Neher, 2002; Sudhof, 2013; Voets, 2000). Following exocytosis, secreted vesicle component are retrieved back into cells by endocytosis. Vesicle trafficking processes are highly dynamic, and different approaches have been developed to study the role of $PI(4,5)P_2$ in vesicle trafficking.

Accumulated work suggests multiple roles of $PI(4,5)P_2$ in regulating different steps of vesicle trafficking. Biochemistry work has identified a phosphatidylinositol transfer protein and a type I PIP5 kinase during vesicle secretion (Hay et al., 1995; Hay and Martin, 1993). Genetic knockouts of synaptojanin-1 (Cremona et al., 1999) and PIP kinase type 1γ (PIPK1γ) (Di Paolo et al., 2004) impair clathrin-mediated endocytosis (CME), vesicle uncoating (Cremona et al., 1999), and the readily-releasable pool (RRP) size (Di Paolo et al., 2004). Over-expression of membrane-targeted synaptojanin-1 and knockdown of PIPK1γ in chromaffin cells decreases the RRP size and vesicle refilling rate (Milosevic et al., 2005), implying a defect upstream the Ca^{2+} triggering step. However, PIPK1γ KO in chromaffin cells showed a selective defect in vesicle priming without affecting vesicle docking and Ca^{2+} currents (Gong et al., 2005). It is noteworthy that most previous work reported in literature employs either *in vitro* assays or cell-wide $PI(4,5)P_2$ perturbations. It remains uncertain to what extent the conclusions from these studies can be applied to *in vivo*, where fast, localized $PI(4,5)P_2$ alterations are often required for signaling specificity and efficiency. The lack of methods for acute and local $PI(4,5)P_2$ manipulations in living cells is a big challenge. Traditional experiments in the field rely on pharmacological or genetic perturbations of key enzymes for PI metabolism, in which global perturbations can evoke non-specific signaling and thus complicate data interpretations. Novel approaches are necessary to better define $PI(4,5)P_2$ function in the context of cell physiology. Recent technology development overcomes the limits and shows promise in tackling this issue. Chemical-inducible approaches, such as rapamycin-induced FRB/FKBP12 dimerization (Hammond et al., 2012; Heo et al., 2006; Suh et al., 2006; Szentpetery et al., 2010; Varnai et al., 2006), can rapidly control $PI(4,5)P_2$ signaling in living cells. Moreover, light-inducible methods provide another new strategy for interrogating cell signaling in time and space (Pathak et al., 2013; Toettcher et al., 2011). Multiple light-inducible systems have been developed in live cells based on natural or engineered photoreceptors, such as light-inducible cryptochrome 2 (CRY2)/CIB1 dimers from *Arabidopsis thaliana* (Idevall-Hagren et al., 2012; Kennedy et al., 2010) and light-oxygen-voltage (LOV) domains (Yazawa et al., 2009). Using motor proteins (kinesin or dynein) in live cells (Duan et al., 2015; van Bergeijk et al., 2015), light-inducible organelle transportation has been demonstrated.

Building on this progresses, a novel optogenetic approach has been developed to control the local $PI(4,5)P_2$ levels in time in living cells. This approach, in combination with total internal reflection fluorescence microscopy (TIRFM), allows examining the acute role of the local $PI(4,5)P_2$ in different steps of vesicle trafficking (Ji et al., 2017). The experiments show that the exact role of $PI(4,5)P_2$ in vesicle trafficking depends on its spatial and temporal changes. An acute and global $PI(4,5)P_2$ reduction strongly inhibits secretion and intracellular Ca^{2+} ($[Ca^{2+}]_i$) responses, but it does not affect sytaxin1a clustering. However, the rapid $PI(4,5)P_2$ reduction locally at the vesicle-PM contacting sites leads to strong vesicle undocking from the plasma membrane, suggesting its crucial role in stabilizing vesicle tethering and docking. This work reveals a previously unappreciated effect of $PI(4,5)P_2$ at the upstream of vesicle priming in the secretory pathway, in addition to its role in vesicle priming and fusion. This effect has a direct physiological relevance since fast and localized $PI(4,5)P_2$ metabolism occurs frequently *in vivo*.

USING SUPER RESOLUTION IMAGING TO STUDY THE MOLECULAR SPATIAL ORGANIZATION OF PIS IN THE PLASMA MEMBRANE

The plasma membrane of eukaryotic cells has a complex topographic organization. The organization pattern of diverse protein clusters or macro-protein complexes in confined membrane area (Engelman, 2005; Spira et al., 2012) facilitates the efficiency and specificity of cellular signaling. For example, syntaxin1A can form numerous dense clusters in the PM (Barg et al., 2010; Ji et al., 2015; Knowles et al., 2010; Sieber et al., 2007), which are believed to be important for membrane fusion. Lipid rafts contain a complex of highly concentrated cholesterol, Shingolipid, and proteins (Lingwood and Simons, 2010), which are thought to be key platforms for cellular signaling. Whether the similar clustering-organization applies to phosphoinositides (PIs) in the PM is debatable. Different results are reported, depending on the approaches, cell types, and experimental conditions. They range from uniform distribution (Hammond et al., 2009; Milosevic et al., 2005; van Rheenen and Jalink, 2002), large patches (Huang et al., 2004; James et al., 2008), to small dense puncta (Aoyagi et al., 2005; James et al., 2008; Kabachinski et al., 2014; Laux et al., 2000; Wang and Richards, 2012).

PI(4,5)P_2 can act as a recruiter or cofactor of proteins for their membrane targeting and translocation. It directly interacts with these proteins at their structured domains, such as pleckstrin homology (PH) domain, ENTH/ANTH, FERM, FYVE, PX, and TARF domains (Hammond and Balla, 2015). The negative charges of PI(4,5)P_2 (-3 to -5) also promote its interaction with proteins containing polybasic stretches of amino acids, such as myristoylated alanine-rich C kinase substrate (MARCKS) and GAP43 (McLaughlin and Murray, 2005; Murray et al., 1999). The PH domain of PLCδ1 has a high affinity in binding PI(4,5)P_2 (Kd = 2 μM) (Garcia et al., 1995). This property allows using $PH_{PLC\delta1}$-tagged fluorescence proteins as a PI(4,5)P_2 probe to visualize PI(4,5)P_2 dynamics and distribution in living cells. Similar fluorescent proteins based on the specific PH domains for different PI species have been broadly used to study PI signaling (Hammond and Balla, 2015).

Mechanistic understanding of PI(4,5)P_2 signaling requires the spatial information of PI(4,5)P_2 molecular organization at the nanometer scale. However, addressing this question is challenging because of the limitation of available methods, as well as the rapid dynamics and equilibrium of PI network in live cells (Ji et al., 2015). The spatial resolution of conventional light microscopy is limited by light diffraction (~250 nm). Electron microscopy (EM) has a nanometer resolution but cannot be applied to live cells, and its harsh sample process raises a concern of distorting PI(4,5)P_2

organization *in vivo* (Ji et al., 2015). Fluorescence resonance energy transfer (FRET) is another method to study protein interactions within ~10 nm but is invalid when the distance is larger than that.

Applications of super-resolution optical imaging shed new lights on this question. A study using stimulated-emission depletion (STED) microscopy (Hell and Wichmann, 1994) reveals abundant dense $PI(4,5)P_2$ clusters (~73 nm in diameter) in PC-12 cell membrane sheets (van den Bogaart et al., 2011). It is proposed that such $PI(4,5)P_2$ dense clusters are required for syntaxin1A clustering through ionic protein-lipid interactions, although weak syntaxin1A self-association can also produce syntaxin1A clustering (Sieber et al., 2007). On the other hand, a study using rapid freezing EM (Fujita et al., 2009; van Rheenen et al., 2005), an approach that preserves the intact PM structure of live cells, observed no dense $PI(4,5)P_2$ nanodomains. Instead, different microdomains of $PI(4,5)P_2$ are observed adjacent to endocytosis sites. Clearly, a coherence conclusion has not reached on this question, as it does by using conventional microscopy.

Single-molecule localization microscopy (SMLM) achieves up to a several-nm spatial resolution and is promising for macro-complex study. It accurately identifies the locations of individual molecules using a point-spread function if they are imaged at well-separated locations and times. Based on this principle, different SMLM approaches have been developed with some modifications of imaging methods. They include photoactivated localization microscopy (PALM) (Betzig et al., 2006), fluorescence photoactivation localization microscopy (FPALM) (Hess et al., 2006), stochastic optical reconstruction microscopy (STORM) (Rust et al., 2006; Xu et al., 2013), ground state depletion microscopy followed by individual molecule return (Folling et al., 2008), and direct STORM (dSTORM) (Wombacher et al., 2010). Among these methods, PALM uses photo-activatable fluorescent proteins (PA-FP) and has advantages of high specificity of labeling by genetically tagging PA-FP to the protein of interest (Betzig et al., 2006; Huang et al., 2013; Pertsinidis et al., 2013; Sengupta et al., 2012; Subach et al., 2009; Szymborska et al., 2013; Zhang et al., 2012).

Using SMLM, the nanoscale organization of $PI(4,5)P_2$ and other key phosphoinositides in the PM have been imaged in INS-1 cells, an insulin-secreting cell line (Ji and Lou, 2016; Ji et al., 2015). This work uncovers an overall homogeneous distribution of $PI(4,5)P_2$ probes in the major PM areas in both fixed and live cells. However, PM micro-domains/patches with enriched PI(4,5)P_2 probes have also been observed (e.g., at cell periphery) although their concentration gradients are relatively low. They display variable sizes and diverse shapes, which are comparable to those observed in the quick-freezing study (Fujita et al., 2009). In addition, PI4P and PI(3,4,5)P_3 show similar spatial patterns as PI(4,5)P_2. This study reveals a detailed nanoscale landscape of PIs in the native plasma membrane. The low concentration gradient (rather than dense clusters) of PIs is likely the primary nanoscale organization form in the PM, and this molecular organization may be sufficient to coordinate diverse PI signaling in living cells.

ACKNOWLEDGMENTS

This work is supported by NIH grants R01DK093953 (to X.L.) and NIH R21NS101584 (to X.L.).

REFERENCES

Abe, N., Inoue, T., Galvez, T., Klein, L. & Meyer, T. (2008). Dissecting the role of PtdIns(4,5)P2 in endocytosis and recycling of the transferrin receptor. *J Cell Sci*, *121*, 1488-1494.

Aikawa, Y. & Martin, T. F. (2003). ARF6 regulates a plasma membrane pool of phosphatidylinositol(4,5)bisphosphate required for regulated exocytosis. *J Cell Biol*, *162*, 647-659.

Aoyagi, K., Sugaya, T., Umeda, M., Yamamoto, S., Terakawa, S. & Takahashi, M. (2005). The activation of exocytotic sites by the formation of phosphatidylinositol 4,5-bisphosphate microdomains at syntaxin clusters. *J Biol Chem*, *280*, 17346-17352.

Arac, D., Chen, X., Khant, H. A., Ubach, J., Ludtke, S. J., Kikkawa, M., Johnson, A. E., Chiu, W., Sudhof, T. C. & Rizo, J. (2006). Close membrane-membrane proximity induced by Ca(2+)-dependent multivalent binding of synaptotagmin-1 to phospholipids. *Nat Struct Mol Biol*, *13*, 209-217.

Arai, Y., Ijuin, T., Takenawa, T., Becker, L. E. & Takashima, S. (2002). Excessive expression of synaptojanin in brains with Down syndrome. *Brain Dev*, *24*, 67-72.

Bai, J., Tucker, W. C. & Chapman, E. R. (2004). PIP2 increases the speed of response of synaptotagmin and steers its membrane-penetration activity toward the plasma membrane. *Nat Struct Mol Biol*, *11*, 36-44.

Balla, T. (2013). Phosphoinositides: tiny lipids with giant impact on cell regulation. *Physiol Rev*, *93*, 1019-1137.

Balla, T., Szentpetery, Z. & Kim, Y. J. (2009). Phosphoinositide signaling: new tools and insights. *Physiology (Bethesda)*, *24*, 231-244.

Barg, S., Knowles, M. K., Chen, X., Midorikawa, M. & Almers, W. (2010). Syntaxin clusters assemble reversibly at sites of secretory granules in live cells. *Proc Natl Acad Sci U S A*, *107*, 20804-20809.

Barlow, C. A., Laishram, R. S. & Anderson, R. A. C. P. C. N. (2010). Nuclear phosphoinositides: a signaling enigma wrapped in a compartmental conundrum. *Trends Cell Biol*, *20*, 25-35.

Berberian, K., Torres, A. J., Fang, Q., Kisler, K. & Lindau, M. (2009). F-actin and myosin II accelerate catecholamine release from chromaffin granules. *J Neurosci*, *29*, 863-870.

Berman, D. E., Dall'Armi, C., Voronov, S. V., McIntire, L. B., Zhang, H., Moore, A. Z., Staniszewski, A., Arancio, O., Kim, T. W. & Di Paolo, G. (2008). Oligomeric amyloid-beta peptide disrupts phosphatidylinositol-4,5-bisphosphate metabolism. *Nat Neurosci*, *11*, 547-554.

Bethoney, K. A., King, M. C., Hinshaw, J. E., Ostap, E. M. & Lemmon, M. A. (2009). A possible effector role for the pleckstrin homology (PH) domain of dynamin. *Proc Natl Acad Sci U S A*, *106*, 13359-13364.

Betzig, E., Patterson, G. H., Sougrat, R., Lindwasser, O. W., Olenych, S., Bonifacino, J. S., Davidson, M. W., Lippincott-Schwartz, J. & Hess, H. F. (2006). Imaging intracellular fluorescent proteins at nanometer resolution. *Science*, *313*, 1642-1645.

Bhandari, R., Juluri, K. R., Resnick, A. C. & Snyder, S. H. (2008). Gene deletion of inositol hexakisphosphate kinase 1 reveals inositol pyrophosphate regulation of insulin secretion, growth, and spermiogenesis. *Proc Natl Acad Sci U S A*, *105*, 2349-2353.

Brose, N. & Rosenmund, C. (2002). Move over protein kinase C, you've got company: alternative cellular effectors of diacylglycerol and phorbol esters. *J Cell Sci*, *115*, 4399-4411.

Chang-Ileto, B., Frere, S. G., Chan, R. B., Voronov, S. V., Roux, A. & Di Paolo, G. (2011). Synaptojanin 1-mediated PI(4,5)P2 hydrolysis is modulated by membrane curvature and facilitates membrane fission. *Dev Cell*, *20*, 206-218.

Chapman, E. R. (2008). How does synaptotagmin trigger neurotransmitter release? *Annu Rev Biochem*, *77*, 615-641.

Cremona, O., Di Paolo, G., Wenk, M. R., Luthi, A., Kim, W. T., Takei, K., Daniell, L., Nemoto, Y., Shears, S. B., Flavell, R. A., et al., (1999). Essential role of phosphoinositide metabolism in synaptic vesicle recycling. *Cell*, *99*, 179-188.

Czech, M. P. (2003). Dynamics of phosphoinositides in membrane retrieval and insertion. *Annu Rev Physiol*, *65*, 791-815.

De Camilli, P., Emr, S. D., McPherson, P. S. & Novick, P. (1996). Phosphoinositides as regulators in membrane traffic. *Science*, *271*, 1533-1539.

Di Paolo, G. & De Camilli, P. (2006). Phosphoinositides in cell regulation and membrane dynamics. *Nature*, *443*, 651-657.

Di Paolo, G. & Kim, T. W. (2011). Linking lipids to Alzheimer's disease: cholesterol and beyond. *Nat Rev Neurosci*, *12*, 284-296.

Di Paolo, G., Moskowitz, H. S., Gipson, K., Wenk, M. R., Voronov, S., Obayashi, M., Flavell, R., Fitzsimonds, R. M., Ryan, T. A. & De Camilli, P. (2004). Impaired PtdIns(4,5)P2 synthesis in nerve terminals produces defects in synaptic vesicle trafficking. *Nature*, *431*, 415-422.

Dominguez, V., Raimondi, C., Somanath, S., Bugliani, M., Loder, M. K., Edling, C. E., Divecha, N., da Silva-Xavier, G., Marselli, L., Persaud, S. J., et al., (2011). Class II phosphoinositide 3-kinase regulates exocytosis of insulin granules in pancreatic beta cells. *J Biol Chem*, *286*, 4216-4225.

Duan, L., Che, D., Zhang, K., Ong, Q., Guo, S. & Cui, B. C. P. C. N. (2015). Optogenetic control of molecular motors and organelle distributions in cells. *Chem Biol*, *22*, 671-682.

Engelman, D. M. (2005). Membranes are more mosaic than fluid. *Nature*, *438*, 578-580.

Falasca, M., Hughes, W. E., Dominguez, V., Sala, G., Fostira, F., Fang, M. Q., Cazzolli, R., Shepherd, P. R., James, D. E. & Maffucci, T. (2007). The role of phosphoinositide 3-kinase C2alpha in insulin signaling. *J Biol Chem*, *282*, 28226-28236.

Folling, J., Bossi, M., Bock, H., Medda, R., Wurm, C. A., Hein, B., Jakobs, S., Eggeling, C. & Hell, S. W. (2008). Fluorescence nanoscopy by ground-state depletion and single-molecule return. *Nat Methods*, *5*, 943-945.

Ford, M. G., Pearse, B. M., Higgins, M. K., Vallis, Y., Owen, D. J., Gibson, A., Hopkins, C. R., Evans, P. R. & McMahon, H. T. (2001). Simultaneous binding of PtdIns(4,5)P2 and clathrin by AP180 in the nucleation of clathrin lattices on membranes. *Science*, *291*, 1051-1055.

Frere, S. G., Chang-Ileto, B. & Di Paolo, G. (2012). Role of phosphoinositides at the neuronal synapse. *Subcell Biochem*, *59*, 131-175.

Fujita, A., Cheng, J., Tauchi-Sato, K., Takenawa, T. & Fujimoto, T. (2009). A distinct pool of phosphatidylinositol 4,5-bisphosphate in caveolae revealed by a nanoscale labeling technique. *Proc Natl Acad Sci U S A*, *106*, 9256-9261.

Gaidarov, I. & Keen, J. H. (1999). Phosphoinositide-AP-2 interactions required for targeting to plasma membrane clathrin-coated pits. *J Cell Biol*, *146*, 755-764.

Garcia, P., Gupta, R., Shah, S., Morris, A. J., Rudge, S. A., Scarlata, S., Petrova, V., McLaughlin, S. & Rebecchi, M. J. (1995). The pleckstrin homology domain of phospholipase C-delta 1 binds with high affinity to phosphatidylinositol 4,5-bisphosphate in bilayer membranes. *Biochemistry*, *34*, 16228-16234.

Gong, L. W., Di Paolo, G., Diaz, E., Cestra, G., Diaz, M. E., Lindau, M., De Camilli, P. & Toomre, D. (2005). Phosphatidylinositol phosphate kinase type I gamma regulates dynamics of large dense-core vesicle fusion. *Proc Natl Acad Sci U S A*, *102*, 5204-5209.

Grishanin, R. N., Kowalchyk, J. A., Klenchin, V. A., Ann, K., Earles, C. A., Chapman, E. R., Gerona, R. R. & Martin, T. F. (2004). CAPS acts at a prefusion step in dense-core vesicle exocytosis as a PIP2 binding protein. *Neuron*, *43*, 551-562.

Halstead, J. R., Jalink, K. & Divecha, N. (2005). An emerging role for PtdIns(4,5)P2-mediated signalling in human disease. *Trends Pharmacol Sci*, *26*, 654-660.

Hammond, G., Schiavo, G. & Irvine, R. (2009). Immunocytochemical techniques reveal multiple, distinct cellular pools of PtdIns4P and PtdIns (4, 5) P2. *Biochem J*, *422*, 23-35.

Hammond, G. R. & Balla, T. (2015). Polyphosphoinositide binding domains: Key to inositol lipid biology. *Biochim Biophys Acta*, *1851*, 746-758.

Hammond, G. R. V., Fischer, M. J., Anderson, K. E., Holdich, J., Koteci, A., Balla, T. & Irvine, R. F. (2012). PI4P And PI(4,5)P(2) Are Essential But Independent Lipid Determinants Of Membrane Identity. *Science* (New York, NY), *337*, 727-730.

Han, X., Wang, C. T., Bai, J., Chapman, E. R. & Jackson, M. B. (2004). Transmembrane segments of syntaxin line the fusion pore of Ca2+-triggered exocytosis. *Science*, *304*, 289-292.

Harris, T. W., Hartwieg, E., Horvitz, H. R. & Jorgensen, E. M. (2000). Mutations in synaptojanin disrupt synaptic vesicle recycling. *J Cell Biol*, *150*, 589-600.

Hay, J. C., Fisette, P. L., Jenkins, G. H., Fukami, K., Takenawa, T., Anderson, R. A. & Martin, T. F. (1995). ATP-dependent inositide phosphorylation required for Ca(2+)-activated secretion. *Nature*, *374*, 173-177.

Hay, J. C. & Martin, T. F. (1993). Phosphatidylinositol transfer protein required for ATP-dependent priming of Ca(2+)-activated secretion. *Nature*, *366*, 572-575.

Hell, S. W. & Wichmann, J. (1994). Breaking the diffraction resolution limit by stimulated emission: stimulated-emission-depletion fluorescence microscopy. *Opt Lett*, *19*, 780-782.

Heo, W. D., Inoue, T., Park, W. S., Kim, M. L., Park, B. O., Wandless, T. J. & Meyer, T. (2006). PI(3,4,5)P₃ and PI(4,5)P₂ Lipids Target Proteins with Polybasic Clusters to the Plasma Membrane. *Science*, *314*, 1458-1461.

Hess, S. T., Girirajan, T. P. & Mason, M. D. (2006). Ultra-high resolution imaging by fluorescence photoactivation localization microscopy. *Biophys J*, *91*, 4258-4272.

Holz, R. W., Hlubek, M. D., Sorensen, S. D., Fisher, S. K., Balla, T., Ozaki, S., Prestwich, G. D., Stuenkel, E. L. & Bittner, M. A. (2000). A pleckstrin homology domain specific for phosphatidylinositol 4, 5-bisphosphate (PtdIns-4,5-P2) and fused to green fluorescent protein identifies plasma membrane PtdIns-4,5-P2 as being important in exocytosis. *J Biol Chem*, *275*, 17878-17885.

Hoy, M., Berggren, P. O. & Gromada, J. (2003). Involvement of protein kinase C-epsilon in inositol hexakisphosphate-induced exocytosis in mouse pancreatic beta-cells. *J Biol Chem*, *278*, 35168-35171.

Huang, F., Hartwich, T. M., Rivera-Molina, F. E., Lin, Y., Duim, W. C., Long, J. J., Uchil, P. D., Myers, J. R., Baird, M. A., Mothes, W., et al., (2013). Video-rate nanoscopy using sCMOS camera-specific single-molecule localization algorithms. *Nat Methods*, *10*, 653-658.

Huang, P., Yeku, O., Zong, H., Tsang, P., Su, W., Yu, X., Teng, S., Osisami, M., Kanaho, Y., Pessin, J. E., et al., (2011). Phosphatidylinositol-4-phosphate-5-kinase alpha deficiency alters dynamics of glucose-stimulated insulin release to improve glucohomeostasis and decrease obesity in mice. *Diabetes*, *60*, 454-463.

Huang, S., Lifshitz, L., Patki-Kamath, V., Tuft, R., Fogarty, K. & Czech, M. P. (2004). Phosphatidylinositol-4,5-bisphosphate-rich plasma membrane patches organize active zones of endocytosis and ruffling in cultured adipocytes. *Mol Cell Biol*, *24*, 9102-9123.

Idevall-Hagren, O., Dickson, E. J., Hille, B., Toomre, D. K. & De Camilli, P. (2012). Optogenetic control of phosphoinositide metabolism. *Proc Natl Acad Sci U S A*, *109*, E2316-2323.

Illies, C., Gromada, J., Fiume, R., Leibiger, B., Yu, J., Juhl, K., Yang, S. N., Barma, D. K., Falck, J. R., Saiardi, A., et al., (2007). Requirement of inositol pyrophosphates for full exocytotic capacity in pancreatic beta cells. *Science*, *318*, 1299-1302.

Imig, C., Min, S. W., Krinner, S., Arancillo, M., Rosenmund, C., Sudhof, T. C., Rhee, J., Brose, N. & Cooper, B. H. (2014). The morphological and molecular nature of synaptic vesicle priming at presynaptic active zones. *Neuron*, *84*, 416-431.

Itoh, T., Koshiba, S., Kigawa, T., Kikuchi, A., Yokoyama, S. & Takenawa, T. (2001). Role of the ENTH domain in phosphatidylinositol-4,5-bisphosphate binding and endocytosis. *Science*, *291*, 1047-1051.

Jahn, R. & Scheller, R. H. (2006). SNAREs--engines for membrane fusion. *Nat Rev Mol Cell Biol*, *7*, 631-643.

James, D. J., Khodthong, C., Kowalchyk, J. A. & Martin, T. F. (2008). Phosphatidylinositol 4,5-bisphosphate regulates SNARE-dependent membrane fusion. *J Cell Biol*, *182*, 355-366.

Ji, C., Fan, F. & Lou, X. (2017). Vesicle Docking Is a Key Target of Local PI(4,5)P2 Metabolism in the Secretory Pathway of INS-1 Cells. *Cell Rep*, *20*, 1409-1421.

Ji, C. & Lou, X. (2016). Single-molecule Super-resolution Imaging of Phosphatidylinositol 4,5-bisphosphate in the Plasma Membrane with Novel Fluorescent Probes. *J Vis Exp*.

Ji, C., Zhang, Y., Xu, P., Xu, T. & Lou, X. (2015). Nanoscale Landscape of Phosphoinositides Revealed by the Specific PH-domains Using Single-molecule Super-resolution Imaging in the Plasma Membrane. *J Biol Chem*.

Kabachinski, G., Yamaga, M., Kielar-Grevstad, D. M., Bruinsma, S. & Martin, T. F. (2014). CAPS and Munc13 utilize distinct PIP2-linked mechanisms to promote vesicle exocytosis. *Mol Biol Cell*, *25*, 508-521.

Kennedy, M. J., Hughes, R. M., Peteya, L. A., Schwartz, J. W., Ehlers, M. D. & Tucker, C. L. (2010). Rapid blue-light-mediated induction of protein interactions in living cells. *Nat Methods*, *7*, 973-975.

Knowles, M. K., Barg, S., Wan, L., Midorikawa, M., Chen, X. & Almers, W. (2010). Single secretory granules of live cells recruit syntaxin-1 and synaptosomal associated protein 25 (SNAP-25) in large copy numbers. *Proc Natl Acad Sci U S A*, *107*, 20810-20815.

Lam, A. D., Tryoen-Toth, P., Tsai, B., Vitale, N. & Stuenkel, E. L. (2008). SNARE-catalyzed fusion events are regulated by Syntaxin1A-lipid interactions. *Mol Biol Cell*, *19*, 485-497.

Landman, N., Jeong, S. Y., Shin, S. Y., Voronov, S. V., Serban, G., Kang, M. S., Park, M. K., Di Paolo, G., Chung, S. & Kim, T. W. (2006). Presenilin mutations linked to familial Alzheimer's disease cause an imbalance in phosphatidylinositol 4,5-bisphosphate metabolism. *Proc Natl Acad Sci U S A*, *103*, 19524-19529.

Laux, T., Fukami, K., Thelen, M., Golub, T., Frey, D. & Caroni, P. (2000). GAP43, MARCKS, and CAP23 modulate PI(4,5)P(2) at plasmalemmal rafts, and regulate cell cortex actin dynamics through a common mechanism. *J Cell Biol*, *149*, 1455-1472.

Lawrence, J. T. & Birnbaum, M. J. (2003). ADP-ribosylation factor 6 regulates insulin secretion through plasma membrane phosphatidylinositol 4,5-bisphosphate. *Proc Natl Acad Sci U S A*, *100*, 13320-13325.

Liao, H., Ellena, J., Liu, L., Szabo, G., Cafiso, D. & Castle, D. (2007). Secretory carrier membrane protein SCAMP2 and phosphatidylinositol 4,5-bisphosphate interactions in the regulation of dense core vesicle exocytosis. *Biochemistry*, *46*, 10909-10920.

Lingwood, D. & Simons, K. (2010). Lipid rafts as a membrane-organizing principle. *Science*, *327*, 46-50.

Lou, X., Korogod, N., Brose, N. & Schneggenburger, R. (2008). Phorbol esters modulate spontaneous and Ca2+-evoked transmitter release via acting on both Munc13 and protein kinase C. *J Neurosci*, *28*, 8257-8267.

Lou, X., Scheuss, V. & Schneggenburger, R. (2005). Allosteric modulation of the presynaptic Ca2+ sensor for vesicle fusion. *Nature*, *435*, 497-501.

Loyet, K. M., Kowalchyk, J. A., Chaudhary, A., Chen, J., Prestwich, G. D. & Martin, T. F. (1998). Specific binding of phosphatidylinositol 4,5-bisphosphate to calcium-dependent activator protein for secretion (CAPS), a potential phosphoinositide effector protein for regulated exocytosis. *J Biol Chem*, *273*, 8337-8343.

Majerus, P. W. & York, J. D. (2009). Phosphoinositide phosphatases and disease. *J Lipid Res*, *50 Suppl*, S249-254.

Mani, M., Lee, S. Y., Lucast, L., Cremona, O., Di Paolo, G., De Camilli, P. & Ryan, T. A. (2007). The dual phosphatase activity of synaptojanin1 is required for both efficient synaptic vesicle endocytosis and reavailability at nerve terminals. *Neuron*, *56*, 1004-1018.

Martin, T. F. (2001). PI(4,5)P(2) regulation of surface membrane traffic. *Curr Opin Cell Biol*, *13*, 493-499.

Martin, T. F. (2015). PI(4,5)P-binding effector proteins for vesicle exocytosis. *Biochim Biophys Acta*, *1851*, 785-793.

McCrea, H. J. & De Camilli, P. (2009). Mutations in phosphoinositide metabolizing enzymes and human disease. *Physiology* (Bethesda), *24*, 8-16.

McLaughlin, S. & Murray, D. (2005). Plasma membrane phosphoinositide organization by protein electrostatics. *Nature*, *438*, 605-611.

Meunier, F. A., Osborne, S. L., Hammond, G. R., Cooke, F. T., Parker, P. J., Domin, J. & Schiavo, G. (2005). Phosphatidylinositol 3-kinase C2alpha is essential for ATP-dependent priming of neurosecretory granule exocytosis. *Mol Biol Cell*, *16*, 4841-4851.

Milosevic, I., Giovedi, S., Lou, X., Raimondi, A., Collesi, C., Shen, H., Paradise, S., O'Toole, E., Ferguson, S., Cremona, O., et al., (2011). Recruitment of endophilin to clathrin-coated pit necks is required for efficient vesicle uncoating after fission. *Neuron*, *72*, 587-601.

Milosevic, I., Sorensen, J. B., Lang, T., Krauss, M., Nagy, G., Haucke, V., Jahn, R. & Neher, E. (2005). Plasmalemmal phosphatidylinositol-4,5-

bisphosphate level regulates the releasable vesicle pool size in chromaffin cells. *J Neurosci*, *25*, 2557-2565.

Murray, D., Arbuzova, A., Hangyas-Mihalyne, G., Gambhir, A., Ben-Tal, N., Honig, B. & McLaughlin, S. (1999). Electrostatic properties of membranes containing acidic lipids and adsorbed basic peptides: theory and experiment. *Biophys J*, *77*, 3176-3188.

Murray, D. H. & Tamm, L. K. (2009). Clustering of syntaxin-1A in model membranes is modulated by phosphatidylinositol 4,5-bisphosphate and cholesterol. *Biochemistry*, *48*, 4617-4625.

Nakano-Kobayashi, A., Yamazaki, M., Unoki, T., Hongu, T., Murata, C., Taguchi, R., Katada, T., Frohman, M. A., Yokozeki, T. & Kanaho, Y. (2007). Role of activation of PIP5Kgamma661 by AP-2 complex in synaptic vesicle endocytosis. *EMBO J*, *26*, 1105-1116.

Neher, E. & Sakaba, T. (2008). Multiple roles of calcium ions in the regulation of neurotransmitter release. *Neuron*, *59*, 861-872.

Olsen, H. L., Hoy, M., Zhang, W., Bertorello, A. M., Bokvist, K., Capito, K., Efanov, A. M., Meister, B., Thams, P., Yang, S. N., et al., (2003). Phosphatidylinositol 4-kinase serves as a metabolic sensor and regulates priming of secretory granules in pancreatic beta cells. *Proc Natl Acad Sci U S A*, *100*, 5187-5192.

Osborne, S. L., Wen, P. J., Boucheron, C., Nguyen, H. N., Hayakawa, M., Kaizawa, H., Parker, P. J., Vitale, N. & Meunier, F. A. (2008). PIKfyve negatively regulates exocytosis in neurosecretory cells. *J Biol Chem*, *283*, 2804-2813.

Pang, Z. P., Cao, P., Xu, W. & Sudhof, T. C. (2010). Calmodulin controls synaptic strength via presynaptic activation of calmodulin kinase II. *J Neurosci*, *30*, 4132-4142.

Pathak, G. P., Vrana, J. D. & Tucker, C. L. C. P. C. N. (2013). Optogenetic control of cell function using engineered photoreceptors. *Biol Cell*, *105*, 59-72.

Perera, R. M., Zoncu, R., Lucast, L., De Camilli, P. & Toomre, D. (2006). Two synaptojanin 1 isoforms are recruited to clathrin-coated pits at different stages. *Proc Natl Acad Sci U S A*, *103*, 19332-19337.

Pertsinidis, A., Mukherjee, K., Sharma, M., Pang, Z. P., Park, S. R., Zhang, Y., Brunger, A. T., Sudhof, T. C. & Chu, S. (2013). Ultrahigh-resolution imaging reveals formation of neuronal SNARE/Munc18 complexes in situ. *Proc Natl Acad Sci U S A*, *110*, E2812-2820.

Posor, Y., Eichhorn-Grunig, M. & Haucke, V. (2015). Phosphoinositides in endocytosis. *Biochim Biophys Acta*, *1851*, 794-804.

Rettig, J. & Neher, E. (2002). Emerging roles of presynaptic proteins in Ca++-triggered exocytosis. *Science*, *298*, 781-785.

Rickman, C., Medine, C. N., Dun, A. R., Moulton, D. J., Mandula, O., Halemani, N. D., Rizzoli, S. O., Chamberlain, L. H. & Duncan, R. R. (2010). t-SNARE protein conformations patterned by the lipid microenvironment. *J Biol Chem*, *285*, 13535-13541.

Rust, M. J., Bates, M. & Zhuang, X. (2006). Sub-diffraction-limit imaging by stochastic optical reconstruction microscopy (STORM). *Nat Methods*, *3*, 793-795.

Saraswati, S., Adolfsen, B. & Littleton, J. T. (2007). Characterization of the role of the Synaptotagmin family as calcium sensors in facilitation and asynchronous neurotransmitter release. *Proc Natl Acad Sci U S A*, *104*, 14122-14127.

Schuske, K. R., Richmond, J. E., Matthies, D. S., Davis, W. S., Runz, S., Rube, D. A., van der Bliek, A. M. & Jorgensen, E. M. (2003). Endophilin is required for synaptic vesicle endocytosis by localizing synaptojanin. *Neuron*, *40*, 749-762.

Sengupta, P., Van Engelenburg, S. & Lippincott-Schwartz, J. (2012). Visualizing cell structure and function with point-localization super resolution imaging. *Dev Cell*, *23*, 1092-1102.

Sieber, J. J., Willig, K. I., Kutzner, C., Gerding-Reimers, C., Harke, B., Donnert, G., Rammner, B., Eggeling, C., Hell, S. W., Grubmuller, H., et al., (2007). Anatomy and dynamics of a supramolecular membrane protein cluster. *Science*, *317*, 1072-1076.

Spira, F., Mueller, N. S., Beck, G., von Olshausen, P., Beig, J. & Wedlich-Soldner, R. (2012). Patchwork organization of the yeast plasma membrane into numerous coexisting domains. *Nat Cell Biol*, *14*, 640-648.

Stein, A., Radhakrishnan, A., Riedel, D., Fasshauer, D. & Jahn, R. (2007). Synaptotagmin activates membrane fusion through a Ca2+-dependent trans interaction with phospholipids. *Nat Struct Mol Biol*, *14*, 904-911.

Subach, F. V., Patterson, G. H., Manley, S., Gillette, J. M., Lippincott-Schwartz, J. & Verkhusha, V. V. (2009). Photoactivatable mCherry for high-resolution two-color fluorescence microscopy. *Nat Methods*, *6*, 153-159.

Sudhof, T. C. (2013). Neurotransmitter release: the last millisecond in the life of a synaptic vesicle. *Neuron*, *80*, 675-690.

Suh, B. C., Inoue, T., Meyer, T. & Hille, B. (2006). Rapid Chemically Induced Changes of PtdIns(4,5)P₂ Gate KCNQ Ion Channels. *Science*, *314*, 1454-1457.

Suh, B. C. & Hille, B. (2008a). PIP2 is a necessary cofactor for ion channel function: how and why? *Annu Rev Biophys*, *37*, 175-195.

Suh, B. C. & Hille, B. C. P. C. N. (2008b). PIP2 is a necessary cofactor for ion channel function: how and why? *Annu Rev Biophys*, *37*, 175-195.

Suh, B. C., Leal, K. & Hille, B. (2010). Modulation of high-voltage activated Ca(2+) channels by membrane phosphatidylinositol 4,5-bisphosphate. *Neuron*, *67*, 224-238.

Szentpetery, Z., Varnai, P. & Balla, T. C. P. (2010). Acute manipulation of Golgi phosphoinositides to assess their importance in cellular trafficking and signaling. *Proc Natl Acad Sci U S A*, *107*, 8225-8230.

Szymborska, A., de Marco, A., Daigle, N., Cordes, V. C., Briggs, J. A. & Ellenberg, J. (2013). Nuclear pore scaffold structure analyzed by super-resolution microscopy and particle averaging. *Science*, *341*, 655-658.

Toettcher, J. E., Voigt, C. A., Weiner, O. D. & Lim, W. A. C. P. C. N. (2011). The promise of optogenetics in cell biology: interrogating molecular circuits in space and time. *Nat Methods*, *8*, 35-38.

van Bergeijk, P., Adrian, M., Hoogenraad, C. C. & Kapitein, L. C. C. P. C. E. (2015). Optogenetic control of organelle transport and positioning. *Nature*, *518*, 111-114.

van den Bogaart, G., Meyenberg, K., Risselada, H. J., Amin, H., Willig, K. I., Hubrich, B. E., Dier, M., Hell, S. W., Grubmuller, H., Diederichsen,

U., et al., (2011). Membrane protein sequestering by ionic protein-lipid interactions. *Nature*, *479*, 552-555.

van Rheenen, J., Achame, E. M., Janssen, H., Calafat, J. & Jalink, K. (2005). PIP2 signaling in lipid domains: a critical re-evaluation. *Embo J*, *24*, 1664-1673.

van Rheenen, J. & Jalink, K. (2002). Agonist-induced PIP(2) hydrolysis inhibits cortical actin dynamics: regulation at a global but not at a micrometer scale. *Mol Biol Cell*, *13*, 3257-3267.

Varnai, P., Thyagarajan, B., Rohacs, T. & Balla, T. (2006). Rapidly inducible changes in phosphatidylinositol 4, 5-bisphosphate levels influence multiple regulatory functions of the lipid in intact living cells. *The Journal of cell biology*, *175*, 377-382.

Vicinanza, M., D'Angelo, G., Di Campli, A. & De Matteis, M. A. (2008). Function and dysfunction of the PI system in membrane trafficking. *EMBO J*, *27*, 2457-2470.

Viiri, K., Maki, M. & Lohi, O. (2012). Phosphoinositides as regulators of protein-chromatin interactions. *Sci Signal*, *5*, pe19.

Voets, T. (2000). Dissection of three Ca2+-dependent steps leading to secretion in chromaffin cells from mouse adrenal slices. *Neuron*, *28*, 537-545.

Voets, T., Moser, T., Lund, P. E., Chow, R. H., Geppert, M., Sudhof, T. C., & Neher, E. (2001). Intracellular calcium dependence of large dense-core vesicle exocytosis in the absence of synaptotagmin I. *Proc Natl Acad Sci U S A*, *98*, 11680-11685.

Voronov, S. V., Frere, S. G., Giovedi, S., Pollina, E. A., Borel, C., Zhang, H., Schmidt, C., Akeson, E. C., Wenk, M. R., Cimasoni, L., et al., (2008). Synaptojanin 1-linked phosphoinositide dyshomeostasis and cognitive deficits in mouse models of Down's syndrome. *Proc Natl Acad Sci U S A*, *105*, 9415-9420.

Wang, J. & Richards, D. A. (2012). Segregation of PIP2 and PIP3 into distinct nanoscale regions within the plasma membrane. *Biol Open*, *1*, 857-862.

Waselle, L., Gerona, R. R., Vitale, N., Martin, T. F., Bader, M. F. & Regazzi, R. (2005). Role of phosphoinositide signaling in the control of insulin exocytosis. *Mol Endocrinol*, *19*, 3097-3106.

Wen, P. J., Osborne, S. L., Zanin, M., Low, P. C., Wang, H. T., Schoenwaelder, S. M., Jackson, S. P., Wedlich-Soldner, R., Vanhaesebroeck, B., Keating, D. J., et al., (2011). Phosphatidylinositol (4,5)bisphosphate coordinates actin-mediated mobilization and translocation of secretory vesicles to the plasma membrane of chromaffin cells. *Nat Commun*, *2*, 491.

Wenk, M. R., Pellegrini, L., Klenchin, V. A., Di Paolo, G., Chang, S., Daniell, L., Arioka, M., Martin, T. F. & De Camilli, P. (2001). PIP kinase Igamma is the major PI(4,5)P(2) synthesizing enzyme at the synapse. *Neuron*, *32*, 79-88.

Wierda, K. D., Toonen, R. F., de Wit, H., Brussaard, A. B. & Verhage, M. (2007). Interdependence of PKC-dependent and PKC-independent pathways for presynaptic plasticity. *Neuron*, *54*, 275-290.

Williams, D., Vicogne, J., Zaitseva, I., McLaughlin, S. & Pessin, J. E. (2009). Evidence that electrostatic interactions between vesicle-associated membrane protein 2 and acidic phospholipids may modulate the fusion of transport vesicles with the plasma membrane. *Mol Biol Cell*, *20*, 4910-4919.

Wojcik, S. M. & Brose, N. (2007). Regulation of membrane fusion in synaptic excitation-secretion coupling: speed and accuracy matter. *Neuron*, *55*, 11-24.

Wombacher, R., Heidbreder, M., van de Linde, S., Sheetz, M. P., Heilemann, M., Cornish, V. W. & Sauer, M. (2010). Live-cell super-resolution imaging with trimethoprim conjugates. *Nat Methods*, *7*, 717-719.

Xu, K., Zhong, G. & Zhuang, X. (2013). Actin, spectrin, and associated proteins form a periodic cytoskeletal structure in axons. *Science*, *339*, 452-456.

Yazawa, M., Sadaghiani, A. M., Hsueh, B. & Dolmetsch, R. E. (2009). Induction of protein-protein interactions in live cells using light. *Nat Biotechnol*, *27*, 941-945.

Yin, H. L. & Janmey, P. A. (2003). Phosphoinositide regulation of the actin cytoskeleton. *Annu Rev Physiol*, *65*, 761-789.

Yoshihara, M., Guan, Z. & Littleton, J. T. (2010). Differential regulation of synchronous versus asynchronous neurotransmitter release by the C2 domains of synaptotagmin 1. *Proc Natl Acad Sci U S A*, *107*, 14869-14874.

Zhang, M., Chang, H., Zhang, Y., Yu, J., Wu, L., Ji, W., Chen, J., Liu, B., Lu, J., Liu, Y., et al., (2012). Rational design of true monomeric and bright photoactivatable fluorescent proteins. *Nat Methods*, *9*, 727-729.

Zoncu, R., Perera, R. M., Sebastian, R., Nakatsu, F., Chen, H., Balla, T., Ayala, G., Toomre, D. & De Camilli, P. V. (2007). Loss of endocytic clathrin-coated pits upon acute depletion of phosphatidylinositol 4,5-bisphosphate. *Proc Natl Acad Sci U S A*, *104*, 3793-3798.

In: Protein-Lipid Interactions
Editor: Angel Catalá

ISBN: 978-1-53613-125-3

Chapter 6

PROTEIN-LIPID INTERPLAY IN VITAL BIOLOGICAL FUNCTIONS

***Konstantin V. Pavlov*[1], *Sergey A. Akimov*[2,3], *Oleg V. Batishchev*[2,4], *Ksenia V. Chekashkina*[1,2], *Pavel V. Bashkirov*[1,2] and *Eduard V. Bocharov*[4,5,6,*]**

[1]Federal Clinical Center of Physical-Chemical Medicine of FMBA, Moscow, Russian Federation
[2]Frumkin Institute of Physical Chemistry and Electrochemistry RAS, Moscow, Russian Federation
[3]National University of Science and Technology "MISiS," Moscow, Russian Federation
[4]Moscow Institute of Physics and Technology, Dolgoprudnyi, Russian Federation
[5]National Research Centre "Kurchatov Institute," Moscow, Russian Federation
[6]Shemyakin–Ovchinnikov Institute of Bioorganic Chemistry RAS, Moscow, Russian Federation

[*] Corresponding Author: E. V. Bocharov. Email: bon@nmr.ru

ABSTRACT

Proteins and lipids play substantially different but complementary roles in plasma membranes. Proteins composed of polymeric chains of chemically and physically diverse amino acid residues capable of adopting versatile and interchangeable spatial configurations combining highly ordered, anisotropically flexible and intrinsically disordered regions are ideal molecular machines sustaining nearly every process in living cells. Virtually unlimited variability of proteins allows them to adapt to any specific function in an evolutionary manner. Membrane lipids are much less versatile and have a rather narrow range of molecular weights. Their recognized primary role in cellular membranes is that of structural elements performing the barrier function that allows cells to maintain concentration gradients and electrochemical potential differences between the cell interior and exterior, as well as providing a 2D fluid environment enabling membrane protein sorting and lateral mobility, determining their folding and spatial organization. However, such a paradigm, in which the lipid bilayer is merely a passive venue for various biological functions performed by highly specialized protein molecules, appears to be an oversimplification. An essential prerequisite of life is dynamic behavior, manifested in conformational transitions of proteins and rearrangements (demixing, changes of packaging order) of the lipid matrix. Intrinsic coupling appears to exist between these processes, and the ability of the lipid subsystem to separate into coexisting liquid phases factors into this coupling in a multifaceted and dynamic manner.

In a rapidly growing number of known examples, biologically significant functions appear to be mediated or driven by the dynamic processes occurring in lipid bilayers. Thus, in the process of membrane tube fission by dynamin protein, the dynamin proved to act as a mere catalyst of the physico-chemical transition resulting in a change of the membrane topology. Some cell signaling mechanisms were shown to rely heavily on the involvement of lipids in the signal transduction cascade, e.g., through specific interactions of certain lipid species with proteins, or through directed changes of physical properties that can propagate laterally to support functional coordination with other proteins and across the membrane to mediate allosteric changes in its opposite leaflet. Various protein-lipid interactions unite lipid matrix and individual membrane proteins into an integral self-consistent plasma membrane system. Selective interactions of proteins with specific lipids, the easiest kind to experimentally resolve and analyze for biological implications, are broadly recognized and characterized rather well. However, increasingly compelling and diverse lines of experimental and physiological evidence corroborated by theoretical analysis and *in silico* modeling suggest existence and biological relevance of higher-order interactions and their ability to fill the gaps in our current understanding of complex and finely regulated behavior of biological membranes. Examples of the driving forces for such regulation include hydrophobic and hydrophilic interactions, mechanical deformation and curvature mediated interactions, concordant changes of protein and lipid properties induced by change of temperature or electric field distribution, and phase-separation coupled interactions. These phenomena suggest several functional and regulatory mechanisms often disregarded in the design and analysis of *in vitro* experiments and interpretation of biological and pathophysiological findings. In addition to the mechanisms supported by direct experimental evidence and detailed calculations, some more speculative options based on plausible inferences are discussed in this review, the distinction between the two being explicitly maintained. As can be seen from the examples analyzed in the review, only a few out of the plethora of possible interaction modes and physical mechanisms appear to be essential in the case of each particular protein or its biological function, justifying the hope that manageable experimental systems and theoretical models can be designed to put these hypotheses to definitive tests.

1. INTRODUCTION

Living cells are self-organizing, open, dissipative systems maintained in a thermodynamically non-equilibrium state at the expense of the continuous inflow of matter and negative entropy from without counteracted by the continuous outbound flow of dissipating energy and life products. The plasma membranes of cells, as well as the membranes of intracellular organelles, provide a fundamental prerequisite for long-term existence of such stationary non-equilibrium conditions. They are often described as selective barriers protecting the low-entropy interior of cells from the chaos and perils of the wide world beyond. However, the expression "barrier function" can be misleading when applied to plasma membranes, whose role is not so much to separate, as to couple the processes and parameters of the cell interior and exterior through orchestrated action of a broad array of membrane-associated protein machines. Life is essentially reliant on the ability of cells not only to exchange matter and energy with the environment across the cellular boundary. Besides that, in order to survive the cells need to receive and process information about changing environmental parameters and respond with flexibility and precision to a broad spectrum of external stimuli, including mechanical, electrical, chemical, and biochemical, as well as to generate and communicate such stimuli. All the vital functions, including the rates of metabolic processes, expression of different factors and morphogenesis of cells, have to be precisely tuned in accordance with the incoming signals, and the information about current status of cells needs to be communicated to their counterparts, especially in the case of multicellular organisms.

Rapid development of structural and analytical tools and methods in the recent decades has greatly expanded our understanding of the processes associated with biological functions of protein machinery on the molecular and submolecular levels. For practical reasons, so far the wealth of the accumulated information on the atomistic details of structure and functioning of water-soluble proteins has greatly surpassed the available data on integral and peripheral proteins. This, however, should not be construed as being representative of either the relative importance attributed to these two categories or of the amount of efforts spent on their investigation. Such an imbalance contributes to continued popularity of the aforementioned tempting paradigm, in which plasma membrane is seen as a mere barrier separating interior of living cells from environment, protecting the cell against natural dissipative processes and providing anchorage for proteins responsible for various interfacial and intercellular communication functions.

2. Membrane-Associated Proteins – Abundance, Biological Functions, and Interplay

Based on different estimates, α-helical transmembrane (TM) proteins alone account for 15 to 40% of organism's proteome (36% based on human serum proteome analysis (Dan et al., 2013)). Taking into account the peripheral membrane proteins that were recently shown to occupy a previously unsuspected 20-25% of the expressed proteome (e.g., in E.Coli, (Monje-Galvan and Klauda, 2016)), and fleeting interactions of the cytoplasmic proteins capable of avidly binding the membrane under appropriate conditions, it appears that we are only beginning to appreciate the scale, role, and complexity of interactions of the protein machinery with surfaces and interiors of biological membranes. The processes induced, essentially controlled, or driven by such interactions go far beyond cellular homeostasis and selective transport of matter across cell boundary, covering full spectrum of essential biological functions, including energy/entropy interconversions, biosynthesis, primary sensor functions (cell signaling and environment probing), information transfer and processing (through lipid-environment-dependent cross-talk between different receptor classes), morphogenesis and life cycle regulation (proliferation, apoptosis, necrosis), etc. Specialized proteins or groups of proteins are credited to be responsible for each of these processes, however, it is becoming increasingly clear that many of the proteins are very sensitive to dynamically changing local lipid environment and can lose, gain or change functions in response to subtle adjustments of the composition or physical state of the surrounding lipids. Diverse and independent lines of evidence suggest that the robust and resilient protein machinery is reliant upon and finely regulated by the chemical composition and physical state of lipids in direct contact with, or in the vicinity of, the proteins involved. Functional importance and versatility of protein-lipid interactions can also be deduced from the number of lipid species found in biological membranes and high degree of conservativeness of the membrane lipid compositions inherent to different organelles, cells or tissues. The biologically relevant lipid species vary in many respects, having different polar headgroups (charge, size, H-bonding properties), hydrophobic tails (number of tails, chain length, saturation, branching, condensed cycles), geometry (head to tail specific area ratio), gel-fluid phase transition temperatures (in homoiothermals the membrane lipid composition is relatively close to the phase transition temperature, rendering membrane the ability to rapidly and coherently respond to temperature variations). Employing such a variety of structural materials to construct a wall with adjustable penetrations for instrumentation and controls would be too wasteful for the evolution to tolerate.

A non-exhaustive list of biological functions performed within, on the surface of, by, or with essential participation of cellular membranes includes locomotion and chemotaxis, transmission, amplification and processing of electric signals in neuronal networks, bone and tissue morphogenesis, cell-to-cell communication mediated by allosteric regulation of various enzymes and transporters, biosynthesis and energy/entropy interconversions, selective transport of ions and metabolites across and along the boundaries of cells and organelles, and immobilization and selective proper presentation of ligands and substrates for membrane-associated enzymes and receptors on the membrane surface (2D catalysis). Transition from 3D volume to 2D surface topology greatly optimizes pre-exponential factor of the rate of the reaction between the membrane-embedded enzymes or receptors and water-soluble ligands. Not only the Brownian route to the reaction center is shorter in 2D space, but the ligand orientation in this case is fixed or strongly restricted, especially in the vicinity of a target partitioning into liquid-ordered phase with more uniformly oriented lipid heads, increasing the chance of successful interaction on approach. The concept is illustrated on Figure 1A by the example of nicotinic acetylcholine receptor (nAChR) inhibition by snake neurotoxin II, where such a mechanism was explored to explain the unusually fast action of the toxin (Lesovoy et al., 2009; Krabben et al., 2009). Some of these functions, e.g., those related to metabolism and homeostasis, are continuously performed, while others, such as cell signaling, fission, or fusion, are occasional and need to be triggered by certain events in the cell life cycle or significant change in cellular environment parameters, or, as is the case with the functions related to ovum fertilization, apoptosis, necrosis and other cell-death scenarios, can only be executed once in the cell lifetime. Another distinction of interest for the purpose of this review is that while some functions are binary ("yes or no") processes or responses to stimuli, in other cases there are multiple levels or diverse discrete outcomes selected by the cell based on additional factors, not all of which can be traced to interactions with additional ligands or regulatory proteins. This is the case with many cell-signaling pathways, where receptors are often capable of triggering differentiated cellular responses to different stimuli (Saito et al., 2004), or, as the case may be, to the same stimulus depending on the initial conditions, including the functional state of other proteins, signaling or otherwise. E.g., functional correlation and a system of feedbacks appears to exist between the epidermal growth factor receptor (EGFR) signaling and the functions of the sodium-potassium transporter ATPase (Na,K-ATPase) (Barwe et al., 2005; Lee et al., 2015). It is tempting, and often possible, to attribute such orchestrated signaling to competitive ligand binding (Doerner et al., 2015), interactions with regulatory ligands and/or direct protein-protein interactions (Rahman et al., 2015; Chaplin et al., 2017), but some cases appear to defy such straightforward explanations and require other physico-chemical mechanisms to be employed. A change of interaction with the surrounding lipids modulated directly by the external stimuli or by protein rearrangements induced by them is the most logical, and in some cases arguably the only viable candidate for such a mechanism. Ability of cells to communicate effectively with each other and their environment relies on simultaneous interpretation and analysis of multiple inputs from different receptors.

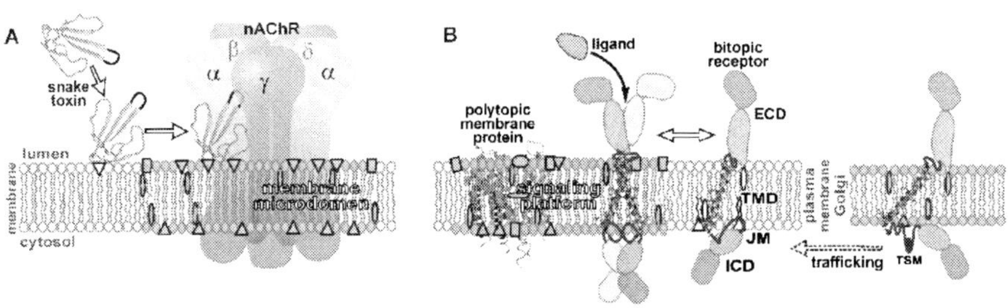

Figure 1. Illustration of lateral protein-protein interactions forced by lipid bilayer. A, "Membrane catalysis" for the nicotinic acetylcholine receptor (nAChR) inhibition pathway of snake neurotoxin II (Lesovoy et al., 2009). Step 1: unspecific electrostatic adsorption of the positively charged toxin from the solution to the negatively charged surface of a cell membrane near the membrane microdomain (raft) surrounding the receptor. Step 2: anchoring of the toxin to negatively charged lipid headgroups (shown by triangles) in a specific orientation that favors the receptor recognition. Step 3: neurotoxin lateral diffusion and subsequent recognition of the binding site between the α and γ subunits of nAChR. B, Functional coupling between the hydrophobic mismatch protein-lipid interaction mechanism and phase transitions between Lo and Ld phases associated with a change of the membrane hydrophobic thickness and implications of hydrophobic mismatch for post-translational protein trafficking. The thinner Golgi membrane (Sharpe et al., 2010) can prevent activation of the transported protein, simultaneously baring the JM region with the specific trafficking motif. Partitioning into the thicker Lo phase of plasma membrane can provide a more precise hydrophobic matching, raft lipid counterparts for specific interactions (headgroups shown by rectangules and triangles) and cholesterol (shown by vertical ovals), and reduce activation entropy because of higher lipid packaging order. Abbreviations: ECD, extracellular domain; TMD, transmembrane domain; ICD, intracellular domain; JM, juxtamembrane region; TSM, trafficking signal motif in cytoplasmic JM region (He et al., 2002; Lemmon and Schlessinger, 2010) of the bitopic protein.

These inputs define life and death choices continuously made by cells, like granting access to a potential nutrient, messenger or intruder, initiating an apoptotic or necrotic pathway, proliferative fission fusion, and any failure to respond to even a weak signal, as well as a spurious actuation of a signaling network, can be the last choice for the cell to make. The required level of complexity of signal processing is achieved through organization of receptors into higher order clusters, often surrounded by the environment of highly specific lipid composition, and the same protein can behave differently in and out of such proteolipidic clusters. This is illustrated by Figure 1B, further discussed in the theoretical sections below in the context of phase separation and mechanisms of the preferential recruitment of proteins into liquid-ordered domains and modulation of the protein conformational landscape by the properties of lipid environment. Another lipid-dependent cross talk scenario is shown of Figure 2A.

3. Lipid-Protein Interaction Role in Bitopic Receptor Signaling – Structural Clues

The need to include into consideration the surrounding lipids as indispensable accomplices of the membrane protein functions appears to be especially compelling for understanding the mechanisms behind the cross-talk and multiple-end-state signaling by bitopic receptor proteins, in which the sensor unit on the extracellular side is linked to the effector inside the cell by a single transmembrane helix, the connection between the extracellular and intracellular sides often being further weakened by flexible or intrinsically disordered juxtamembrane linker segments. These receptors commonly need to form functionally competent dimers *via* interfaces located in the cytoplasmic, intracellular and TM parts. Predictably enough, it is the switch between alternative dimerization interfaces within the TM helices that has been shown to underlie their activation in response to ligand binding in many cases (Landau and Ben-Tal, 2008; Bocharov et al., 2010b; Endres et al., 2013; Bocharov et al., 2013; Sharonov et al., 2014; Waters and Brooks, 2015; Maruyama, 2015; Bocharov et al., 2016; Bragin et al., 2016; Sarabipour and Hristova, 2016; Bocharov et al., 2017a; Bocharov et al., 2017c). The TM dimerization interfaces are formed by groups of amino acid residues that are better off in contact with the each other than exposed to the hydrophobic core of the membrane. From this standpoint, taking into account that the lengths of such interfaces on the TM helices are in the range 7 to 12 residues (Bugge et al., 2016) and the typical length of the entire TM helix is about 25 residues, even one alternative dimerization interface means a significant free energy penalty that has to be compensated by preferential interactions of the remaining part of the helix with hydrocarbon tails of the lipid membrane. This leaves very little room at best for having multiple alternative dimerization sites that could explain how the TM part of the receptor conveys different messages across the membrane. On the other hand, a prototypical member of the receptor tyrosine kinase (RTK) superfamily, EGFR, provides a lucid example of differentiated signaling in which protein-lipid interactions appear to play a vital role. The receptor is capable of binding an array of different ligands, resulting in distinguishable ligand-dependent end states of the cell (Wilson et al., 2009). The signals transduced in response to binding of the seven growth factors known to activate EGFR in mammalian cells fall into three distinct categories, in which the structure of the coiled-coil formed by intracellular juxtamembrane (JM) segment correlates with the downstream signaling outcomes (Doerner et al., 2015). Concordant results of *in vivo*, *in vitro* and *in silico* experiments provide compelling evidence that the identity of the ligand is translated into different temporal dynamics and magnitude of cellular response *via* formation of alternative JM coiled-coil rotamers. Moreover, mutations in the JM-A region were shown to alter the energetics of ligand binding to the ectodomain (ECD) (Macdonald-Oberman and Pike, 2009). The JM-A sequence contains both a hydrophobic helical interface (a common feature of natural water-soluble proteins forming coiled-coil superstructures) and a number of salt bridging residues, making it capable of adopting multiple functionally relevant conformations. This model, however, implies existence of some mechanism propagating the signal across the membrane, and the only link between the ligand-binding ectodomain and the catalytic intracellular part, the α-helical single-span TM region, in itself appears to be lacking flexibility and diversity of allowable conformational states to be up to the task. Besides the physically limited space to accommodate multiple alternative dimerization interfaces (dormant EGFR exists primarily in the dimeric form, so at least one dimerization interface has to be "inactive") in the α-helical TM domain (TMD), there is yet another constraint. The energy barriers between the functionally relevant conformational states have to be high enough to prevent spontaneous activation of a function, but not as high as to interfere with sensitivity to all the stimuli the receptor is intended to respond to. The putative multiple "active" TMD dimer conformations have to be separated from each other and from the "passive" conformation by the (ligand-binding-dependent) energy barriers conformant with these restrictions. In the same time, these different TMD conformation have to be deterministically inducing different conformations of the JM-A segment in order to faithfully translate the identity of a bound ligand into the downstream cellular response. It would appear that having a congruent hierarchy of such conformational states is simply too much to demand from a pair of rigid single-span α-helixes. On the other hand, the membrane lipids form a flexible and dynamic environment with multiple variable properties sensitive to interactions with not only the transmembrane part of the protein, but also with the water-soluble segments that can form transitory contacts with the lipid heads. Moreover, the hydrophobic coiled-coil interface of the flexible intracellular JM-A segment can, and was explicitly demonstrated to be able to reversibly plunge under the lipid head surface (Endres et al., 2013; Mineev et al., 2015). Thus, in the case of EGFR, and RTK in the broader context, the nature of the conformational rearrangements of the protein upon activations appears to suggest the mode of interaction with lipids that might form a missing link in the mini-cascade of events responsible for the protein function. As discussed in the (Bocharov et al., 2016; Bocharov et al., 2017a), dimerization *via* the less polar of the two alternative dimerization interfaces within the TMD in the presumably inactive state of the receptor exposes the more polar interface to contact with the lipids in the hydrophobic core of the membrane. Thermodynamically, it has to induce a local change in the physico-chemical state of the lipid, increasing its dielectric permittivity and permeability to water through a change of order parameters and/or local chemical composition in order to match the properties of the exposed part of the membrane helix.

The dynamic properties of the lipid bilayer allow for propagation of this disturbance of local membrane properties both in the lateral and transversal direction, providing a physical mechanism for allosteric regulation of the state of the intramembrane part of the receptor machinery by ligand binding on the extracellular side. Switching to the more polar TMD dimerization interface attenuates the disturbance and increases the local hydrophobicity, and hence changes the lipid interaction with the TM and JM parts, inclusively in the intracellular side of the membrane. Possibility of this effect having sufficient magnitude to make the mechanism viable is supported by MD simulations and theoretical models based on the mean-field approach (Bocharov et al. 2017b). It was further corroborated by the experimentally observed two-way correlation between the state of the lipid environment and conformation of the transmembrane dimer, when a change of the membrane mimetic (from DMPC/DHPC bicelles to DPC micelles) in the otherwise identical conditions switched the EGFR dimerization mode from the presumably "active" to the presumably "inactive" one. The simplicity of the membrane-mimicking model in this case only emphasizes the fact that a mere and rather subtle change of a scalar physico-chemical property of the lipid environment can trigger a functionally relevant conformational transition of the protein. Interestingly, the hypothetical mechanism of such lipid-mediated EGFR signaling prompted by these observations and analysis of other available structural, biochemical and physiological information involved release from the inner leaflet of the plasma membrane and subsequent dimerization of the amphiphilic flexible intracellular JM-A segments of the protein, in gratifying agreement with the above discussed findings related to the role of the coiled-coil rotamers of the JM-A in the differentiated signaling by the EGFR. Figure 2B illustrates the proposed signaling mechanism in isolation, and Figure 3 below provides a glimpse into the its possible implications for the EGFR cohabitants in the proteolipidic clusters, offering an explanation for graded and diverse signaling outcomes conditioned by parallel processes occurring on the membrane.

This example suggests a certain distribution of roles between the intramolecular and intermolecular protein interactions on one side, and lipid environment rearrangements and protein-lipid interactions on the other, in the execution and regulation of membrane-associated biological functions. Both the lipidic and the protein components of the membrane have a degree of stability needed to support cellular homeostasis and resistance to adverse factors, and ability to react to external or internal stimuli through conformational or phase transitions.

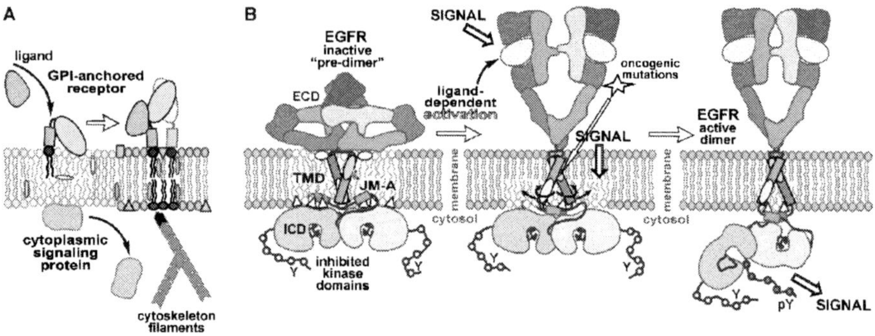

Figure 2. Putative lipid-mediated mechanism of signal transduction across the membrane by GPI-anchored and bitopic membrane proteins assuming explicit role of lipid environment as a part of a self-consistent signaling system. **A**, Transmembrane signaling by a glycosyl-phosphatidyl inositol (GPI)-anchored protein based on local coupling of lipid states between the membrane leaflets and partitioning into a proteolipid domain. A signaling cascade initiated by interaction of a ligand-bound receptor with the outer leaflet conveyed to the inner leaflet through propagation of the lipid state disturbance across the membrane can result in downstream signaling events on the inner surface and in the cytoplasm, e.g., cytoskeleton remodeling (Suzuki, 2015; Sharonov et al., 2016). The acyl chains of lipids in a different ordering (perturbed) state are shown in *lighter shade*, functionally relevant specific lipid headgroups – by triangles, cylinders and *black* ovals, cholesterol molecules are shown by elongated ovals. **B**, Sequence (*left to right*) of structural rearrangements of the human epidermal growth factor receptor (EGFR) and associated perturbations of the lipidic bilayer in the course of ligand-induced receptor activation (Bocharov et al., 2016). The conventions from Figure 1 apply, TMD and JM α-helical regions are shown by the cylinders. *Firstly*, the interaction of the ECD in the inactive tethered state with the membrane perturbs the lipid bilayer, making it locally more polar (shown in *lighter shade*) thus forcing it to favor the alternative dimerization mode (*via* a hydrophobic C-terminal motif, light oval in the lower part of TMD) and submerged position of amphipatic JM-A regions in the inner leaflet (inhibited configuration). *Secondly*, ligand binding causes the ECD to transition into an extended conformation, weakening its interaction with the membrane immediately over the TMD. This induces changes of local properties (including the increased degree of lipid ordering) that can propagate across the membrane. At the same time, these changes can cause the TMD to switch into the alternative N-terminal conformation (*via* a polar N-terminal motif), compatible with the new, more ordered, lipid environment. This effectively amplifies the lipid perturbations sufficiently to induce the switching, whereas sub-threshold perturbations tend to decay, making the signaling process more definitive. *Finally*, the intracellular leaflet of the membrane and parts of the protein assume the active state characterized by release of the JM-A regions from the membrane (and their subsequent coiled-coil interaction) and association of the kinase domain into an asymmetric active dimer. The effects of TMD pathogenic (mostly oncogenic) mutations on gain-of-function EGFR signaling is proposed under the assumption that forced mutation-induced switching to the active dimerization mode of TMD can modify the physical properties of the lipid bilayer with consequential kinase activation in the absence of a ligand. This can be achieved either by introduction of S-S bonds in the TMD locking the dimer in the active conformation, or by amino-acid substitutions increasing polarity of the active dimerization motif (Li and Hristova, 2010; Bocharov et al., 2013; Ou et al., 2017).

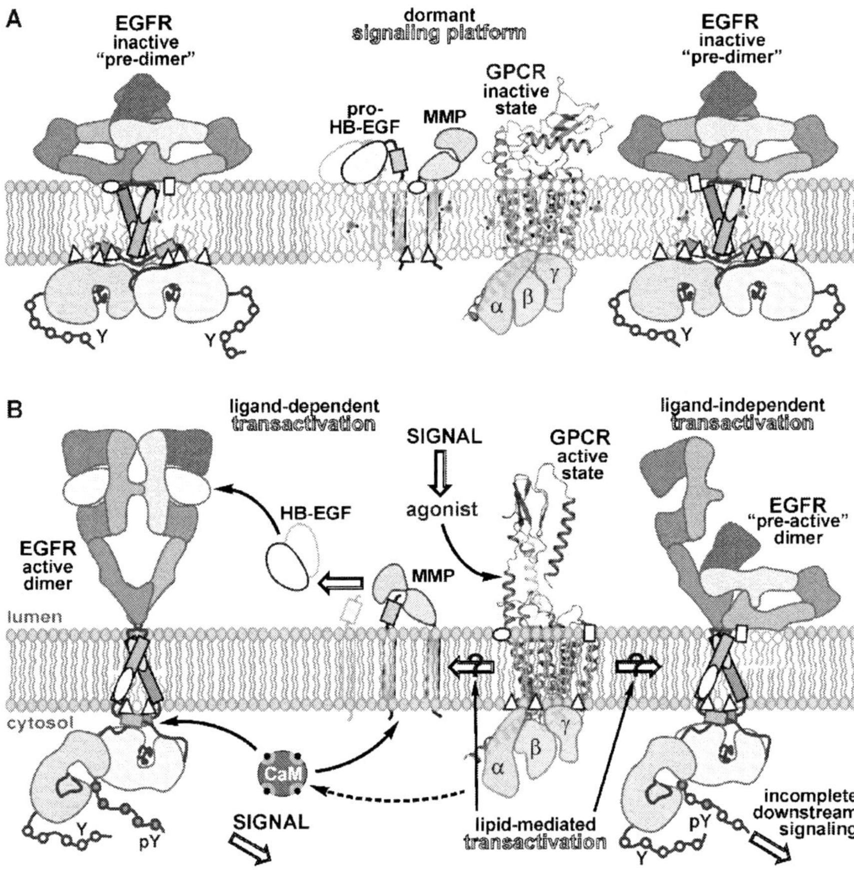

Figure 3. Hypothetic possibilities for the lateral lipid-mediated interplay between the participants of signal transduction in a signaling platform. **A**, "Dormant" signaling platform with embedded G-protein-coupled receptor (GPCR), matrix metalloproteinase (MMP), precursor of heparin-binding epidermal growth factor (pro-HB-EGF) and EGFR. The proteins cooperatively support a "disturbed" state of lipids consistent with their dormant states. The schematically shown water molecules can penetrate into the perturbed membrane more easily. Lipid heads of the specifically interacting lipids are shown by open triangles, rectangles and ovals. **B**, Sequence of structural rearrangements of the membrane proteins induced by ligand-binding to the GPCR and associated evolutions of the lipidic bilayer state and composition, which can be responsible for ligand-independent transactivation of proximal receptors (*right-hand* pathway) resulting in incomplete signaling described in (Chaplin et al., 2017; Köse, 2017), or ligand-dependent activation mediated by, e.g., MMP (*left-hand* pathway). The interplay can be augmented by activation-induced redistribution of specifically interacting rare lipid species and/or by EGFR activity modulation through interaction of Ca^{2+}-bound calmodulin (CaM) with the cytoplasmic JM region of EGFR (Sánchez-González et al., 2010).

However, due to the restraints imposed by the primary sequence of amino acids in the protein, the available conformational space of a polymeric protein molecule with a given number of amino acid residues is much more compact than the conformational (phase) space of 2D fluid surface consisting of an equivalent number of lipid molecules (or having the same total molecular weight). For the same reasons, the energy minima in the protein conformational space should be sharper and further between, and have fewer phase trajectories connecting them without crossing a prohibitively high energy barrier. Therefore, whereas lipid mixtures readily undergo reversible phase transitions, demixing and mixing as the current thermodynamic parameters dictate, an overheated protein denaturates irreversibly. Thus, all the lipids within the same membrane can undergo combinatorial interactions with each other and with non-lipidic components of the membrane, with occasional temporary confinement in liquid-ordered proteolipidic domains (also known as rafts and extensively discussed in the foregoing sections). Even within such domains, lipids enjoy a certain level of freedom to travel, since the domain boundary separating it from the outer fluid phase is never completely impenetrable and the domain is in a dynamic equilibrium with the rest of the membrane. On the other hand, a protein, that can only migrate as a whole, usually escorted by a retinue of specifically bound and annular lipids, is much more restricted in the choice of interaction partners. In summary, while the proteins appear to be much more suitable for ensuring reliability of performance of the membrane-related biological functions and resilience to variable environmental parameters, the lipidic components can add a measure of flexibility of response to the changes and support optimal spatial and temporal organization of diverse protein functions.

4. Specific Lipid-Protein Interactions

The most straightforward examples emphasizing the role of interaction with lipids for membrane-associated protein functions are provided by the proteins requiring direct interaction a specific lipid component, which is often negatively charged and usually rather rare, representing minor fractions in the total lipid composition of the membrane. The most typical examples are gangliosides, sphingomyelin (SM), phosphatidylinositol phosphates (PIPs), cardiolipin (CL), phosphatidic acid (PA), phosphatidylglucerol (PG), or phosphatidylserine (PS). Oftentimes, such interactions can be pinned down to a single function and target an exact binding site, typically presenting a positively charged pocket or surface on a specific protein (Lemmon, 2008), or to even smaller an interacting moiety. For example, the phosphatidylinositol-specific phospholipase C (PI-PLC) from *Bacillus thuringiensis* depend on lipid-specific recognition of target membrane. PI-PLC interacts with PC headgroups *via* cation-π interactions with tyrosine residues, implicating cation-π interactions at the interface in providing a generic mechanism for specific lipid recognition by amphitropic and membrane proteins (Grauffel et al., 2013). E.g., dynamin-related GTPase Grp1 specifically interacts with CL through a four-lysine module in one of its inserts, dramatically enhancing Grp1 olygomerization and GTPase activity (Bustillo-Zabalbeitia et al., 2014). The affinity to other negatively charged lipid species is substantially lower, to the point of undetectable interaction with polyanionic PIPs having higher net negative charge in comparison with CL. PA is a minor phospholipid that plays a central role in several key cellular processes through specific interactions with proteins through diverse PH binding domains containing basic residues. Interactions with the side chains of Lys (K) and Arg (R) side chains are effectively stabilized by hydrogen bonding, turning PA into a preferred docking site for these residues (Kooijman et al., 2007). The structure-activity studies of a member of defensin family, MtDef4, have established that the RGFRRR loop within the protein as a translocation signal mediating entry of this defensin into fungal cells. The RGFRRR loop interacts with the phospholipid phosphatidic acid (PA), and this interaction is presumably essential for the entry of MtDef4 into fungal cells and/or its intracellular toxicity (Sagaram et al., 2013). PIP role in specific interactions with proteins is especially widely recognized. Each cellular membrane compartment contains a characteristic species of phosphoinositide, which attracts a specific complement of functionally important, loosely attached peripheral proteins to that membrane. For example, the phosphatidylinositol 4,5-bisphosphate (PIP_2) of the plasma membrane attracts phospholipase C, protein kinase C, proteins involved in membrane budding and fusion, proteins regulating the actin cytoskeleton, and others. Phosphoinositides also regulate the activity level of the integral membrane proteins. Many ion channels of the plasma membrane also need the plasma-membrane-specific PIP_2 to function. The phosphoinositide regulation of ion channels/transporters has been first clearly postulated in the case of Na^+/Ca^{2+} exchanger (NCX1) and KATP potassium channels studied in giant excised patches of the heart (Hilgemann and Ball, 1996). In summary, phosphoinositides contribute to the selection of peripheral proteins for each membrane and regulate the activity of the integral proteins (Falkenburger et al., 2010). GM1 ganglioside was shown to serve as an intestinal binding receptor for a protein exotoxin produced by the infecting *Vibrio cholera* (Holmgren et al., 1975). Formation of caviolae, bulb-shaped nanodomains of the plasma membrane enriched in cholesterol and sphingolipids and having a broad spectrum of physiological functions, including endocytosis, mechanosensing, and regulation of membrane and lipid transport, mediated by integral membrane proteins caveolins (Cavs) and peripheral proteins cavins proved specifically dependent on plasmalemmal anionic lipid phosphatidylserine (PtdSer) and membrane lipid asymmetry. Both protein families bind anionic phospholipids, however, the functional role of interactions with different negatively charged lipid species proved different. Thus, acute depletion of PI(4)P and PIns(4,5)P_2 had minimal impact on caveola assembly, though diminishing lateral confinement, whereas sequestration or depletion of PtdSer decreased caveolin clustering, co-localization of Cav1 and cavin1, and the number of caveolae, and increased the rate of cavin1 degradation. Moreover, loss of PtdSer asymmetry between the inner and the outer monolayers known to precede initiation of apoptosis and some other bifurcations in the cell life cycle, drastically affected caveolae stability (Hirama et al., 2017).

Most of the lipids discussed on the context of specific protein-lipid recognition are acidic and carry negative charge in the physiological range of pH values. Acidic phospholipids are present in all cell membranes, but the inner leaflet of plasma membrane accommodates a majority of the charged lipid population, distributed heterogeneously and forming micro- and nanodomains under appropriate conditions. In addition to positively charged binding pockets with a well-defined globular structure, they were found to interact with proteins containing polybasic sequences through a slightly different mechanism known as ionic interaction (Heo et al., 2006; Sengupta ett al., 2009; Wang et al., 2014). Unlike the binding sites, polybasic sequences are often unstructured making this interaction mechanism less specific, but giving it an additional regulatory leverage through rearrangements of the flexible polybasic sequence readily induced by the interaction. This feature can be directly relevant to the functional role of the JM-A regions of the EGFR proteins in receptor signaling discussed above. The ionic interactions are extensively involved in regulation of protein structure and function, as is thoroughly reviewed in (Li et al., 2014).

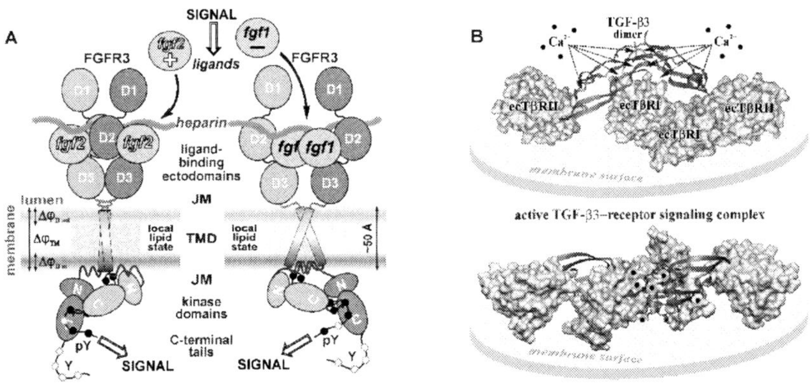

Figure 4. Extended version of the recently proposed lipid-mediated mechanisms of signal transduction across cell membrane augmented by consideration of electrostatic interactions. **A**, Putative structural implications of binding by human FGFR3 receptor of two different ligands, weakly acidic fgf1 and strongly basic fgf2 with substantially different electric charges, resulting in two diverse active configurations of the receptor, according to (Sarabipour and Hristova, 2016; Bocharov et al., 2017c). Besides the difference in conformational states of the ectodomain, lipid-mediated influence of dipole and transmembrane potentials (ΔφD and ΔφTM) on the JM and possibly the TMD regions can be responsible for the differences in the downstream signaling (Kovács et al., 2016), represented by different patterns of solid *black* circles showing phosphorylated tyrosines. **B**, A hypothetical dynamic model for interaction of TGF-β ligands with the receptor dimers and formation of the signaling complex on the cell surface. Formation of the active TGF-β signaling complex after ligand binding to the ectodomains of the type I (TβRI) and type II (TβRII) receptors is decomposed into sequential stages, corresponding to semi-open and extended conformations of the ligand (Bocharov et al., 2002; Bocharov et., 2008). The ligand was shown to have 4 Ca^{2+} binding pockets (8 per dimer) (Bocharov et., 2008), occupancy of which can alter its conformational dynamics, in the same time affecting Coulomb interactions of the free ligand with the receptors (top panel) and the change of membrane electric field distribution induced by the mature signaling complex (bottom panel).

Being very dynamic and dependent on local concentrations of charged proteins, phospholipids and inorganic ions, the ionic protein-lipid interactions are themselves regulated by one of the major intracellular bivalent cations, a key secondary messenger $Ca2^{+}$. This regulatory mechanism is equally applicable to the interactions with specific lipid recognition sites, and considered together with the effects of lipid ordering rearrangements on the dipolar and boundary potential distribution provides a nice finishing touch to the lipid-mediated signaling concept, as illustrated by Figure 4 with the examples of the human fibroblast growth factor receptor (FGFR) differentiated response to ligands with different electric charge, and the dual (though possibly more complex in reality) role of calcium binding by the transforming growth factor-β (TGF-β) ligand. Another hypothetical calcium-sensitive regulatory mechanism based on the competition of the activated CaM with acidic phospholipids for binding to polybasic protein sequences was illustrated in Figure 3B.

It is readily perceivable that specific interactions with individual lipid species or classes of lipids can be beneficial for organization of spatially and temporally correlated work of the membrane-associated protein machinery. Spatial colocation of the functionally related proteins (e.g., into signaling platforms processing sets of inputs that need to be collectively factored into a different set of outcomes) and segregation of those with conflicting functions (e.g., homeostasis and excitability) coupled to preferential partitioning of the lipid species with non-zero spontaneous curvature induced by local membrane geometry does appear to be essential for enabling cells to discharge its multiple concurrent and competitive vital functions in a mutually cognizant but independently adjustable manner. However, multiple experimental facts and theoretical considerations suggest that the interplay between the lipid constituents of the membrane, integral and peripheral proteins and their water-soluble ligands goes beyond such specific, one-to-one and short-range interactions.

5. Non-Specific and Parametric Lipid-Protein Interactions: Pathophysiological Evidence and Paradigmatic Examples

Another lipid often implicated in specific interactions with proteins is cholesterol. Strictly speaking, it does not seem to belong with the category of lipids discussed above: in most of the membranes it is not a minor component, neither is it charged under physiological conditions. Accordingly, the functional role and nature of its interaction with proteins also appears to be different from those described above and belong to a higher hierarchical order. It is lucidly exemplified by a pathological condition known as Smith–Lemli–Opitz syndrome, a multiple malformation syndrome caused by mutation of 7-dehydrocholesterol reductase (DHCR7) impairing the reduction of 7-dehydrocholesterol (7DHC) to cholesterol in the final step of cholesterol biosynthesis. Phenotypically, SLOS ranges from a mild disorder with behavioral and learning problems or autistic characteristics to a lethal developmental dysmorphology. Mutations of DHCR7 result in decreased cholesterol and increased dehydrocholesterol levels (Porter, 2008). 7DHC is normally present in healthy organisms, and normal cholesterol levels do not exclude SLOS. However, SLOS treatment usually involves dietary cholesterol supplementation, alleviating though not curing the disease (DeBarber et al., 2011; Bianconi et al., 2015). In summary, both low cholesterol and elevated 7DHC appear to factor into the condition referred to as SLOS in a complex and cooperative manner. Sterols are essential components of lipid rafts, implicated in many physiological processes affected by this pathology. Theoretical analysis with the aid of elasticity theory and direct experimental observations with phase-separating giant unilamellar vesicles (GUV) revealed, that though replacement of cholesterol for 7DHC does not prevent phase separation and raft formation, the positive spontaneous curvature introduced by the replacement dramatically alters the profile of raft interaction energy. Instead of fusing on contact, as is the case with the model phosphatidylcholine/ sphingomyelin/ cholesterol membranes, in the 7DHC-substituted composition rafts agglutinate into necklace-like aggregates, being unable neither to separate, nor to fuse or exchange matter. Thus, the membranes of the affected cells still contain the domains of liquid-ordered phase known to be important for multiple membrane trafficking, lipid sorting and signaling processes, but the dynamic exchange between rafts and the ability of raft-associated proteins to interact with counterparts confined in other rafts is dramatically limited (Staneva et al., 2016).

Another pathological condition, Gaucher's disease, is also caused by a single dysregulation of metabolism of a single lipid species, namely a genetic deficiency of the lysosomal enzyme glucocerebrozidase (Butters, 2007). Similarly to SLOS, the disease is a polysimptomatic condition, some of the symptoms of which (painless enlargement of spleen and liver, or yellow fatty deposits in the whites of the eye) are readily attributable to accumulation of unmetabolized glucosylceramide, while others (joint and bone malformation, lesions, hypertonia, convulsions, coordination deficiencies, mental retardation or dementia) are more systemic and indicative of dysregulation and imbalance between multiple protein functions rather than loss or constitutional activation of any specific one. Although the symptoms are usually attributed to glucosylceramide storage, the fact that there is little or no correlation between the severity of the symptoms, residual activity of the enzyme and the amount of accumulated lipids casts doubts on the cause and effect relation between the lysosomal lipid accumulation and the disease (Sidransky, 2012). On the other hand, the elevated sphingolipid levels observed in the plasma membrane of cellular models of Gaucher's disease are capable of altering the dipole potential that has its origins in the dipole moments of polar groups of the membrane lipids. Concordantly, the magnitude of the increase of the dipole potential induced by increased sphingolipid concentration in Gaucher disease is similar to the difference between the dipole potential in rafts and in the rest of the membrane estimated in (Kovács et al., 2017) with the aid of several different fluorescent probes. Besides pointing to the possible pathophysiological significance of the increased dipolar potential, it suggests that changes of the microdomain structure of plasma membranes might also play a role in development of some Gausher's symptoms. In a broader context, any change in local or global physico-chemical parameters, such as temperature, boundary, dipolar or transmembrane potential difference, dielectric permittivity, mechanical tension, lipid packaging characteristics (specific area per lipid head, tail orientation, hydrophobic thickness etc.), geometric curvature would modulate the functions of membrane-associated proteins by changing the free energy of interactions with their different functionally relevant conformations, in the same time affecting the phase separation properties of the lipids and hence spatial distributions of proteins. Moreover, for geometrical and thermodynamic reasons all these changes are interdependent, and any theory of lipid-protein interactions inspired or influenced by the mean field theories relating the local orientational order as a function of depth in bilayer membranes to the thermodynamic properties of the liquid-ordered and liquid-disordered phases expressed in terms of a single order parameter can be used to represent any of the various types of interactions considered (Mouritsen and M. Bloom, 1984).

A straightforward and well-understood example of a function controlled by such parametric protein-lipid interactions is thermoreception. In principle, temperature changes can directly induce conformational transitions of proteins, but in the paradigmatic example of the *Bacillus subtilis* thermosensor DesK, the temperature response of the protein appears to be triggered by the temperature-induced changes of the lipid membrane (de Mendoza, 2014). The protein is a polytopic integral histidine kinase, acting as a primary sensor in the unsaturated fatty acid synthesis regulatory pathway. After autophosphorilation of a conserved His residue induced by low temperature, the phosphoryl group is transferred to an aspartic acid residue in the DNA-binding effector DesR downstream of the signaling cascade, activating transcription of the *des* gene encoding the acyl lipid desaturase Δ5-Des (Aguilar et al., 1998). Expression of Δ5-Des upregulates production of unsaturated fatty acids, thus increasing fluidity of the membrane, and eventually terminating the kinase activity of the sensor. Importantly, DesK remains fully functional when reconstituted in pure lipid vesicles, wherefore no other component is involved in either sensing or signaling (Martín et al., 2009). Even more remarkably, a chimeric TM region consisting of N-terminal part of the first TM helix and C-terminal part of the fifth TM helix to test a model based on systematic deletion analysis proved to work almost as well as a wild type DesK with five transmembrane helices. Moreover, two N-terminal hydrophobic amino acid residues (K-10 and N-12) capable of snorkeling across the membrane water interface were identified as primarily responsible for temperature sensing (Cybulski et al., 2010). However, the natural conjecture was the temperature-induced change in the membrane thickness the sensor was responding to; it proved to be an oversimplification. The kinase activity and responsiveness to temperature of the protein reconstituted into lipid bilayers made of different lipids depended non-monotonously on the length of the acyl chains, but correlated with the departure from phase transition temperature of the lipid under the experimental conditions.

6. Lipid Phase Separation of Cell Membrane

One of the vital functions of lipids in the cellular membranes is to provide spatial and temporal organization of the work of protein machinery in the cellular membranes. A typical cell contains several thousands of different membrane proteins and about a hundred of distinct lipid species. Each protein performs or contributes into one or several biological functions, enabled by conformational transitions between a rather limited number of states determined by its primary sequence and native fold provided by specialized machinery, and has certain preferences and requirements to the environment around it. Some of the functions are complementary, while others, such as excitability and homeostasis, endocytosis and exocytosis, are competitive or mutually exclusive. The lipid membrane thus has to ensure sequestration and segregation of the complementary and conflicting functions, respectively, temporary confinement of the proteins performing the functions requiring spatial colocation or strict temporal sequencing, and accessibility of the proteins to the appropriate water-soluble or membrane-bound factors. These functions essentially depend upon the ability of the biological membrane lipids to form coexisting phases, and in our opinion the importance, the broad implications and contentious nature of the phenomena justifies a brief detour into the history behind the current understanding of it. Our vision of the lateral organization of plasma membranes evolved considerably since the advent of the liquid mosaic model (Singer and Nicholson, 1972), which postulated roughly uniform distribution of proteins in the fluid lipid bilayer. EPR measurements of spin label mobility in a model membrane consisting of phospholipids and cytochrome-oxidase (Jost et al., 1973) revealed two states of the label characterized by notably different mobilities, the higher of which corresponded to the fluid membrane. This gave rise to the much debated and controversial concept of annular, or boundary, lipids forming a layer with the thickness of one or two molecules relatively strongly bound to transmembrane proteins. Later (Anderson and Jacobson, 2002), it was proposed that considerably more than one or two layers of the annular lipids can be closely associated with the integral proteins. The so-called lipid shells (see Figure 5) were hypothesized, having the thickness of a few nanometers and the lipidic composition substantially different from the average composition of the plasma membrane. The lipid shells, however, were believed to exist in the same fluid phase state as the entire membrane.

Further experiments demonstrated that plasma membranes can be laterally inhomogeneous, forming different coexistent phases. Solubilization of the membrane with low concentrations of the mild detergent Triton-X100 at the temperature of ~4°C with subsequent centrifugation yielded two fractions, one of which was more detergent resistant, and enriched with a broad spectrum of membrane proteins, sphingomyelin and cholesterol (Edidin, 2001). If cells were treated with high concentrations of the detergent, the fractions of protein, lipid, sphingomyelin and cholesterol separated independently. Similar experiments preceded by cholesterol extraction from the membrane with the aid of β-methylcyclodextrin failed to yield a detergent-resistant fraction. This lead the authors of (Simons and Ikonen, 1997) to the hypothesis of long-term existence of detergent-resistant lipid-protein domains termed “rafts” being in the phase state different from that of the rest of the membrane.

Operational definition of the raft phase as of the fraction obtained from plasma membranes *via* a sequence of manipulations per a certain standard protocol proved rather inconvenient. The lipid and protein composition of the fraction turned out to depend considerably upon the detergent used, its concentration or temperature (García-Marcos et al., 2006). In order to test the raft hypothesis in the natural conditions without preliminarily disintegrating the membrane, the authors of (Pralle et al., 2000) investigated diffusion of latex spheres bound *via* antibodies to membrane proteins anchored to glycosyl-phosphatidyl inositol (GPI), which preferentially partitions into the detergent-resistant phase. The diffusion rate of the proteins was found to be consistent with the assumptions of them being bound to a relatively rigid islet of the lipid bilayer with the effective radius of 26 ± 13 nm. It was later demonstrated by means of transmission electron microscopy that the plasma membrane lipids and proteins form extended aggregates associated with the cytoskeleton, the so-called islands with the characteristic dimension of 200 nm (Lillemeier et al., 2006). Finally, domains were isolated from plasma membrane at physiological temperature and pH without any detergent (Ayuyan and Cohen, 2008). Based on the results of analysis of their lipid and protein compositions, 4 types of domains with different content of membrane proteins (alkaline phosphatase, alkaline diesterase, Na^+-K^+-ATPase) and sphingomyelin were identified, while cholesterol partitioned almost equally into all these domains. It was concluded that the rafts in plasma membranes of cells are relatively small, ranging from 10 to 200 nm, i.e., below the diffraction limit of the visible light microscopy. *In vivo,* investigation of rafts is greatly complicated by continuous recirculation of plasma membranes and functioning of various enzymes, some of which catalyze biochemical interconversions of lipids, while functioning of others affects local physico-chemical properties of the membrane, thus making rafts of cellular membrane highly dynamic and transient structures. For these reasons, even the meaning of the term "raft" remained vague and disputable, until (and arguably even after) the following "consensus definition" was formulated in 2006 by the members of a Keystone Symposium on Lipid Rafts and Cell Function: "Membrane rafts are small (10-200 nm) heterogeneous, highly dynamic sterol and sphingolipid-enriched domains that compartmentalize cellular processes. Small rafts can sometimes be stabilized to form larger platforms through protein-protein and protein-lipid interactions" (Pike, 2006). Even though not all the structures currently associated with rafts meet all the criteria, e.g., the rafts of prokaryotic cells containing no sphingomyelin, this definition is widely accepted as a reference for the identification of raft-like structures.

Simple model membranes containing a saturated acyl chain lipid with high melting temperature (e.g., sphingomyelin), an unsaturated chain lipid with low melting temperature (e.g., DOPC), and cholesterol or other sterol, nucleate micron-sized domains of liquid ordered (L_o) phase enriched in the saturated chain lipid and liquid disordered (L_d) phase enriched in the unsaturated lipid (Samsonov et al., 2001; Veatch et al., 2004). Cholesterol partitions almost equally between the phases, with a slight bias in favor of L_o. The microscopic rafts retain circular shape, which is rapidly (within seconds) restored after external disturbance (Samsonov et al., 2001), which implies considerable line tension of the raft boundary (Baumgart et al., 2003). Based on direct atomic force microscopy (AFM) measurements, rafts are usually 1-2 nm thicker than the lipid around them (Rinia et al., 2001; Galimzyanov et al., 2017b). Multiple lines of evidence suggest that the carbohydrate chains of lipid molecules in the domains are more highly ordered and uniformly oriented, which is usually attributed to interaction with cholesterol. Besides that, on the micron scale rafts in bilayer lipid membranes are almost always in registry (Baumgart et al., 2003; Samsonov et al., 2001; Veatch et al., 2004): if an area of one leaflet of the membrane is in the L_o state, the corresponding area of the opposite monolayer is also liquid ordered. Lateral organization of multicomponent cellular membranes is admittedly far more complex than coexistence of raft and non-raft phases, but barring sufficiently discriminative measurement and modeling techniques, we are still blind to most of the details of such organization.

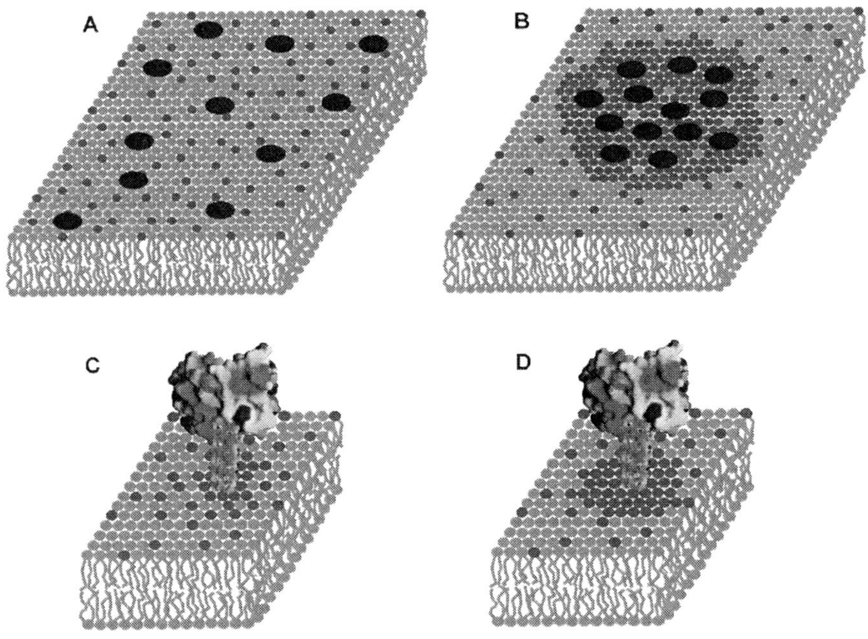

Figure 5. A hierarchy of the low-level lipid-protein structural units in the cellular membrane. (**A**) Fluid mosaic model: homogeneous distribution of lipids and proteins, (**B**) Condensed lipid-protein "islands," (**C**) Boundary (annular) lipids, and (**D**) Lipid shell in the vicinity of an isolated protein (surface representation). Proteins are shown in *black*, lipids – in different shades of *grey*.

7. Mechanisms of Sequestration of Proteins into Ordered Lipid Domains

In vitro experiments with rafts are usually performed at room temperature. For the raftogenic mixture of dioleoylphosphatidylcholine

(DOPC)/dipalmitoylphosphatidylcholine (DPPC)/ cholesterol 1:1:1, optically visible domains disappear at the temperature of about 30°C, i.e., below the physiological temperatures (Veatch et al., 2004). For the membranes formed from lipid extracts of plasma membranes, visible domains disappear at about 25°C (Veatch et al., 2008), though existence of domains smaller than the visible light diffraction limit (e.g., 200 nm) cannot be ruled out based on these results. Förster resonance energy transfer measurements revealed that in the DOPC/SM/Cholesterol 2:1:2 mixture roughly corresponding to the proportions of the major components of cellular membranes, the domains of the size comparable to Förster radius (several nanometers) for nitrobenzoxadiazole and rhodamine pair of dyes disappear around the physiological temperature (Ayuyan and Cohen, 2006; Ayuyan and Cohen, 2008). Thus, it remains unclear whether it is the lipidic or protein component of the cellular membrane that plays a leading role in formation of lipid-protein domains, i.e., does the global phase separation of the lipid subsystem occur first, for the proteins to partition subsequently between the phases, or do proteins nucleate the domains with the lipidic composition of their choosing around them. In any case, for effective lateral sorting purposes, proteins must have high affinity to a certain lipid phase, no matter whether the phase pre-exists in the membrane or is transiently and dynamically condensed around the protein.

Certain proteins and lipids essentially need to cluster in a confined part of the plasma membrane in order to support essential biological processes, including membrane fission and fusion, cell signaling, receptor trafficking etc. Some aspects of ligand-induced receptor clustering and association with signaling proteins are influenced by lipid composition and local physico-chemical properties of membranes. Besides the normal physiological functions, assembly and organization of the protein-lipid platforms underlies many pathological processes, such as assembly of new viral particles on the membranes of cells infected by enveloped viruses. Formation and subsequent budding of an infectious virion requires the entire complement of the viral proteins and genetic materials to be gathered in a confined area of the plasma membrane or in its immediate vicinity. It was experimentally demonstrated that the lipid compositions of the influenza and human immune deficiency viruses are essentially identical to that of the raft phase of the cellular membrane (Polozov et al.,

2008). Assembly of new virions is assumed to take place on the L_o domains, into which the viral proteins have to preferentially partition.

Affinity of various proteins to the L_o phase is sometimes attributed to presence of specific sequences of amino acid residues capable of specific cholesterol recognition (Li and Papadopoulos, 1998), known as "CRAC" (cholesterol recognition amino acid consensus) motifs described by a general formula -L/V-(X)(1-5)-Y-(X)(1-5)-R/K-, where (X)(1-5) stand for one to five residues of arbitrary amino acids. Such sequences were identified in the primary structure of many proteins partitioning preferentially into the raft phase of cellular membranes, including caveolines, multiple receptor proteins, as well as the influenza virus fusion protein hemagglutinin and matrix protein M1 (Li and Papadopoulos, 1998; Epand et al., 2005). Initially the hypothesis about functional role of the CRAC motifs in distribution of proteins between membrane lipid phases was based on the specific "recognition" of cholesterol by the motif. However, cholesterol is experimentally detected in almost invariable concentration throughout the membrane (Ayuyan and Cohen, 2008; Veatch et al., 2004). In the same time, cellular membrane L_o domains are known to be considerably enriched in sphingomyelin (Ayuyan and Cohen, 2008), suggesting an idea that CRAC is more of a misnomer, and the motif can also interact with SM, or that specific interaction with cholesterol depends on presence of SM or other lipid partitioning into rafts as a cofactor. Besides, a cholesterol recognition motif such as so-called GG4-motif in TM helices (two glycine residues spaced by three arbitrary residues) can also act as a protein-protein dimerization motif (MacKenzie, 2006; Bocharov et al., 2010a), creating possibility for competition between dimerization or protein folding and cholesterol binding (Nadezhdin et al., 2011; Nadezhdin et al., 2012; Barrett et al., 2012).

For the peripheral proteins associated with the membrane through a lipidic anchor, or for the proteins covalently bound to hydrophobic tails of lipids during post-translational processing, preferential partitioning into the raft phase can be provided through affinity of their lipid counterpart to this phase. In particular, GPI-anchored proteins are classical markers of the raft phase (Pralle et al., 2000). The influenza virus fusion protein hemagglutinin is known to be covalently bound to three palmitic acid residues during post-translational processing (Chlanda et al., 2017). Taking into account relatively high ordering level of the viral membrane, similar to that of the raft phase of typical cellular plasma membranes (Polozov et al., 2008), it can be concluded that the necessary sequestration of hemagglutinin for assembly of new virus particles is achieved through interactions with saturated palmitoleic chains.

An alternative hypothesis on the mechanism underlying affinity of proteins to different lipid phases is based on the experimental data obtained in the model systems, where the thickness of the ordered raft was shown to be larger than that of the disordered surrounding membrane (Rinia et al., 2001; Galimzyanov et al., 2017b; Bao et al., 2011). Proteins can therefor partition between these phases in accordance with the length of their hydrophobic transmembrane domains. This "hydrophobic mismatch" mechanism, however, is fully applicable only to multiple-span integral proteins and dimers/oligomers with a relatively large monomer crossing angle. Indeed, single-span transmembrane helices or compact dimers with a small distance separating their monomer termini can accommodate to thickermembranes by means of tilting (adjustment of the membrane crossing angle), imposing no (in the case of monomeric) or very little (in the case of a dimer with small crossing angle and hence small distance between the N-terminal and C-terminal ends of the transmembrane helix) additional energy penalty at the edges. Tilting a large "barrel" protein with multiple membrane-spanning chains is impossible without submerging a segment near the edge of its hydrophilic surface into the hydrophobic membrane environment on one side and exposing an equivalent segment of the hydrophobic surface to the aqueous solution on the opposite side near each edge of the transmembrane domain. If the protein is involved in specific interactions with lipids, which usually occur *via* polar head groups, the tilt would also bring the interaction interface out of the phase separation plane and impose an additional energy penalty. It is therefore hardly surprising that no credible correlation was found between the length of the transmembrane domain of a synthetic peptide and the peptide's partitioning between the L_o and L_d phases (McIntosh et al., 2003). However, GPCRs, transport ATPases and other multiple-span proteins and oligomers do not have the orientational flexibility of bitopic proteins, and thus can be very selective to local membrane thickness (Saita and Mendoza, 2015). Indeed, a systematic study of the influence of phosphatidylcholine (PC) acyl chain length and saturation, and of cholesterol concentration upon the catalytic activity, activation energy and reconstitution parameters of purified Na,K-ATPase reconstituted into giant unilamellar vesicles performed in the work (Cornelius, 2001) confirmed crucial importance of hydrophobic matching for the protein function. For functional reconstitution, membrane bound Na,K-ATPase was solubilized with the nonionic detergent $C_{12}E_8$ and mixed with the lipids under investigation solubilized with the same detergent. Then, after excessive mixing time was allowed for equilibrate the composition of the motion-restricted annular lipids with the bulk lipids in the protein-lipid detergent micelles, the detergent was subsequently removed by addition of hydrophobic beads, and large (~100 nm) unilamellar liposomes containing reconstituted the Na,K-ATPase spontaneously formed. For the saturated PC, the protein recovery decreased with increasing acyl chain length, whereas for monounsaturated PC the optimal acyl chain length was 20, and the protein recovery was generally higher. The presence of cholesterol increased the recovery, more effectively so for shorter acyl chains. Selectivity of orientation of the reconstituted ATPase in the membrane also depended notably on the acyl chain length, having a minimum at the chain length of 16. Again, the changes induced by cholesterol were consistently more pronounced for shorter chain lengths. The activity of the Na,K-ATPase reconstituted with the saturated di-C14:0 PC proved higher than with the monounsaturated PC of the same acyl chain length, even at the temperatures below 25 °C where di-C14:0 PC is in the gel phase. This indicates that besides the lipid fluidity, there are other properties of significance for the protein function. The presence of cholesterol shifted the chain length optimum from 22 to 18, equivalent to a shift in hydrophobic thickness from ~34 to ~27 Å, such a decrease being in accordance with what cholesterol is expected to counterbalance by increasing bilayer thickness. Presence of ~40% cholesterol also increased the maximal turnover rate with the PC chain lengths <22, and decreased the activation free energy of the rate limiting step due to entropic contribution, despite the increase of the activation enthalpy it induced. Remarkably, both the turnover rate and the activation free energy in the model liposomes with the optimal PC chain length in the presence of cholesterol were very close to those achieved by the Na,K-ATPase under the native conditions in cellular membranes. Altogether, the observed effects imply that both hydrophobic length and lipid packaging order are important for the reaction catalyzed by the Na,K-ATPase. On the other hand, mere adjustment of these two crucial parameters allowed achieving native or nearly-native performance of the protein under standard conditions, with exception of the regulatory functions. The significance of this result is related to the dawning recognition of the degree individuality of different structural units of heterogeneous cellular membrane and of the broad range of diverse parameters describing this individuality. Possibility of faithfully representing the functionally important properties of the lipid environment for each particular protein with a greatly reduced subset of components and attributes gives ground for cautious optimism with respect to feasibility of description of the system.

Another observation of generic import related to the hydrophobic mismatch mechanism and more broadly the essential properties of TM domains of integral proteins dictated by the lipid environment was made in the work (Sharpe et al., 2010), where a large dataset from fungi and vertebrates were used to perform a comprehensive comparison of the TM domains of bitopic proteins from different organelles. The comprehensive comparison confirmed the previously suggested difference in TMD length and hence bilayer thickness between the TMDs of Golgi and plasma membrane proteins (Bretscher and Munro, 1993; Levine et al., 2000), and revealed an apparent stepwise change in bilayer thickness that occurs in the secretory pathway at the trans- side of the Golgi.

8. Wetting Phenomena

The maximal phase separation temperature of at least some physiologically relevant lipid mixtures in the absence of membrane-associated proteins is below the normal body temperatures of non-hibernating homoiothermals. In the "classical" raft-forming mixture of dioleoylphosphatidylcholine (DOPC)/dipalmitoylphosphatidylcholine (DPPC)/cholesterol 1:1:1, the phase separation temperature is slightly below the physiological one, and in the lipid mixture of DOPC/SM/Cholesterol 2:1:2 nanodomains were shown to disappear at physiological temperatures (Ayuyan and Cohen, 2006; Ayuyan and Cohen, 2008). For the bilayer lipid membranes made of the lipid extract of cellular plasma membranes, the optically visible domains disappear at as low temperature as 25°C (Veatch et al., 2008). Thus, the plasma membranes of homoiothermals do not necessarily exist under the conditions allowing phase separation for the lipid subsystem of the membrane. However, interactions with the membrane- associated proteins can nucleate local phase separation even in the subsaturated lipid mixtures through the phenomena known as "wetting" (Landau and Lifshitz, 2013).

In the absence of the conditions necessary for a global phase transition, the equilibrium state of the membrane is a uniform mixture of lipids and proteins, and any local deviations from this free energy minimum costs the system some work, which has to be compensated by sufficiently strong protein-lipid interactions. The boundaries of the film of a new phase store a certain boundary energy. The specific energy of unit length of the boundary is referred to as line tension. The boundary of the film in the plane of the membrane has a finite curvature radius, creating Laplacian pressure counteracting formation of the wetting film. Thus, for a wetting lipid film to be formed in the under saturated conditions, a certain type of dependence of the film free energy on its thickness is needed in addition to high affinity of lipids to the protein in order to counterbalance the Laplace pressure. In other words, wetting requires not only the "energy balance" to be maintained, but also the "force balance." The specific force equalizing the Laplace pressure (at least for certain values of the film width) is termed "disjoining" or "wedging" pressure, which has to be directed outwards, creating repulsion of the boundary between the film and the bulk membrane and the boundary between the film and the protein.

The wetting phenomenon in relation to membrane proteins and lipids has been so far investigated in details only theoretically (Gil et al., 1997; Gil and Ipsen, 1997; Gil and Mikheev, 1995). Long-wavelength thermal fluctuations of the boundary between the wetting film and the surrounding membrane as a source of wedging pressure were analyzed in a series of works (Gil et al., 1997; Gil and Ipsen, 1997; Gil and Mikheev, 1995). Since this boundary cannot cross the boundary between the film and the protein, effective repulsion of entropic nature occurs between these boundaries. This mechanism of stabilization of the raft phase film is only viable at low line tensions of the boundary between the wetting film and the bulk membrane. Under optimal conditions, the estimated thickness of the film does not exceed 1-2 nm, since the entropic repulsion of the boundaries decays rapidly with the increasing film thickness. However, even such a relatively thin film proves sufficient to induce aggregation of proteins based on the so-called Capillary condensation mechanism (Gil and Ipsen, 1997), associated with formation of the wetting film between the condensing proteins. Thus, the lipid-protein interaction based on the wetting phenomenon can induce effective protein-protein interaction and coordination.

An alternative hypothesis on the nature of the wedging pressure is based on the experimentally observed difference of thickness of the lipid bilayer in the L_o and L_d phases. Proteins are assumed to partition between these phases in accordance with the length of the transmembrane domain: the proteins with the longer TM domains preferentially partition into rafts, while the ones with the shorter TM domains prefer the disordered phase ("hydrophobic matching"). It has been hypothesized (Akimov et al., 2008), that even in the absence of conditions for phase separation of a given lipid mixture, proteins can locally induce formation of a lipid phase around them with the thickness matching their transmembrane domain length. In this case, a stepwise change of the membrane thickness, known as "hydrophobic mismatch," occurs at the boundary between the wetting film and the bulk membrane. If the membrane is assumed to be planar, the energy of a hypothetic boundary with such an abrupt change of membrane thickness would be prohibitively high due to contact of the exposed hydrophobic parts of lipid molecules with water. In order to minimize the free energy (and the boundary line tension), the system needs to minimize the area of contact between the hydrophobic and polar media. In the publication (Akimov et al., 2008) it was assumed to be achieved at the expense of deformation of the membrane in the vicinity of the boundary with a gradual change of the membrane thickness. The energy stored in such mechanical deformation of the bilayer near the boundary defines the line tension in this case. The structure of these deformations turns out to induce repulsion between the film boundaries on the side of the bulk membrane and of the protein, providing the wedging pressure needed to stabilize the film. The theory predicts that macroscopic films with the thickness of about 15 nm can be formed only around large proteins or protein aggregates with the diameter of about 15 nm. Single protein molecules with small TM domains can sustain the wetting films of the thickness of ~4 nm determined by the characteristic lengths of mechanical deformations of the membrane. It was also demonstrated that such a wetting film facilitates entry of individual protein molecules into a protein aggregate surrounded by the film, and complicates withdrawal of single molecules from such an aggregate. Besides that, lipid raft films facilitate fusion between the aggregates.

The discussed mechanisms of occurrence of the wedging pressure are complementary and interrelated. Large hydrophobic mismatch between the raft and the bulk membrane results in high line tension associated with the membrane deformation near the phase boundary. The high line tension inhibits long-wavelength fluctuations of the external film boundary, effectively "canceling" the entropic contribution into the wedging pressure. However, the "deformational" wedging pressure alone is sufficient to stabilize the film. The situation is reversed in the case of small hydrophobic mismatch values, when the energy of deformations needed to compensate it is also small, allowing significant fluctuations of the shape of the outer boundary of the film and increasing relative contribution of the "entropic" wedging pressure.

9. Mechanisms of Coupling between Liquid-Ordered Phase Domains in the Opposing Monolayers of Bilayer Lipid Membranes

The combination of lateral inhomogeneity and phase coexistence with bilayer structure of lipid membranes enables a simple but elegant mechanism that can play a significant role in the cellular signaling processes. As discussed above, proteins incorporated into lipid bilayers are capable of inducing local phase transitions, depending on the type and strength of their interactions with the lipid environment. Thus, conformational transitions induced by receptor-ligand interactions on the extracellular side can cause formation or disappearance of a liquid-ordered phase domain in the outer leaflet of the membrane. Coupling that exists between the phase states of monolayers, depending on the domain size and composition of the domain and of the bulk lipid mixture, implies that a matching L_o domain has to form or disappear in the inner leaflet, thus translating the information about the ligand binding event across the membrane.

Transbilayer dimerization of cholesterol was one of the first mechanisms proposed for the intermonolayer domain coupling (Harris et al., 1995; Rukmini et al., 2001). However, in many mixtures cholesterol partitions almost equally between the phases, and hence such dimerization is insufficient to ensure domain registry. In some theoretical studies, coupling between monolayers is simply postulated (Fowler et al., 2016; Williamson and Olmsted, 2015; Putzel and Schick, 2008) or introduced as a macroscopic parameters obtained from microscopic simulations (Putzel et al., 2011).

Interdigitation of lipid acyl chains from the opposing monolayers was also suggested to underlie the coupling (Slater and Huang, 1988) by conveying the dynamic characteristics from one monolayer to the other (e.g., slowing down thermal movements of lipids), but the mechanism provides no driving force for demixing. SANS ("small angle neutron scattering") method used in (Heberle et al., 2016) to investigate the coupling allowed resolving fluidization of the gel phase caused by contact with a liquid-phase monolayer in asymmetric membranes made of pure DPPC (gel) and POPC-enriched (fluid) monolayers. In the work (Horner et al., 2013), the changes of diffusion rates of fluorescent probes incorporated into charged DOPG membrane caused by unilateral adsorption of polylysine were registered. The changes occurred in both monolayers irrespectively of the carbohydrate chain length of the probe. Comparison of the results of these two publications (Horner et al., 2013; Heberle et al., 2016) suggests that the exact nature of the formation of ordered phase domain in one monolayer is immaterial to the coupling between the monolayers. It can be therefore inferred, that universal physical mechanisms irrespective of the chemical structure of the membrane constituents have to underlie the coupling. The principal parameters differing substantially between the ordered and disordered monolayers are the monolayer thickness and elastic moduli, including the bending modulus. As shown in (Galimzyanov et al., 2015), membrane deformations near the boundary between raft and bulk membrane induced for compensation of the hydrophobic mismatch cause the boundaries of the domains on the opposing monolayers to align almost in registry, with a small relative shift of 2-4 nm. This contribution to the energy of domain coupling depends linearly on the domain size and hence is effective for relatively small domains (< 60 nm). Analysis of the bending fluctuations of the membrane containing liquid disordered parts with low bending modulus, and liquid-ordered domains characterized by high bending rigidity was performed in (Galimzyanov et al., 2017a; Haataja, 2017). In-registry configuration was shown to deliver the energy minimum. The major contribution to the domain coupling in this case comes from short-wave fluctuations insensitive to lateral tension of the membrane or presence of solid support. This contribution is proportional to domain area, i.e., grows quadratically with size and dominates for larger domains (> 60 nm). Thus, between this two mechanisms discussed in (Galimzyanov et al., 2017a; Haataja, 2017; Galimzyanov et al., 2015), coupling between L_o domains in the opposing monolayers is provided in the entire range of their sizes.

10. Linactants and Non-Monotonous Line Tension as a Function of Their Concentrations - Homeostatic Mechanism

Rafts in the cellular plasma membranes are extremely small, different estimates range from 10 to 200 nm (Pralle et al., 2000; Lillemeier et al., 2006). Such a sizes of the domains appear optimal for operation of the membrane proteins, being sufficiently large to accommodate the proteins that need to be collocated in order to perform their intended function, but small enough to prevent collocation of conflicting or incompatible components or network.

Long-term existence of an ensemble of rafts with relatively stable size distribution is yet to find a satisfactory explanation. Thermodynamically, the average domain size in the ensemble is governed by the balance of the domain boundary energy and the translational entropy of the ensemble. The latter benefits from the increased number of domains, and is independent of the membrane phase properties. Above a certain threshold of the boundary line tension, fusion of all domains into a common macroscopic phase minimizes the system free energy, despite the contribution of the associated decrease of the ensemble translational entropy. For sufficiently small line tensions, the translational entropy "wins" dictating unlimited fission of domains. Simple estimates show that in an unconstrained closed system, no thermodynamic equilibrium is possible (there is no free energy minimum) for an ensemble of domains of the same size: they either disintegrate into individual lipid molecules, or merge into one macroscopic domain. The lipid subsystem of the membrane, however, is neither unconstrained nor closed. Cytoskeletal matrix and protein crowding restrict the growth of domain area, and the membrane lipids continuously recirculate and undergo modifications by various enzymes (Gómez et al., 2009), integral and peripheral proteins interact with each other and water-soluble ligands and experience energy-dependent conformational transitions, etc. None of these effects in isolation is sufficient to explain all the experimental observations accumulated so far, but appropriately combined they have provide the required explanation. On the medium time-scale, the nanoraft ensemble can be kinetically trapped in the thermodynamically non-equilibrium state: as demonstrated in (Frolov et al., 2006), domain coalescence ("Ostwald ripening") process can take as long as 10 hours at reasonably low domain boundary line tensions. The definitive role of the line tension for domain size distribution is also corroborated by the results obtained in (Usery et al., 2017), where optically visible domains (>1 micron) were seen to always occur at the same line tension value of ~0.3 pN.

The line tension strongly depends on temperature (Honerkamp-Smith et al., 2008), but the range of physiological temperatures is rather narrow. However, some membrane components, such as GM1 ganglioside, were found to be able to cause surprisingly large changes of the line tension even at tiny concentrations (~0.1 mol%) (Bao et al., 2011). By an analogy with the surfactants in 3D heterogeneous systems, such components of 2D membranes are termed linactants (Brewster and Safran, 2010). The mechanisms of line activity were proposed to be associated with the "hybrid" nature of linactant molecules combining parts with natural preference to different lipid phases (Brewster and Safran, 2010). This is exemplified by POPC, having saturated palmitoyl chain preferring ordered environment, and monounsaturated oleoyl chain with a natural tendency to partition into disordered phase. This mechanism is however contentious, since both phases are liquid and lipid components are continuously in rapid rotational thermal motions (Nojima and Iwata, 2014). The authors of (Galimzyanov et al., 2017b) demonstrated that spontaneous curvature of a lipid is crucial for its linear activity. It is generally determined by the ratio or the effective cross-sections of the lipid head and tail, the convention being that negative curvature is assigned to the "cone" lipids that prefer concave monolayer surfaces, and positive – to "inverse cones" favoring convex monolayers. According to (Galimzyanov et al., 2017b), every component with a sufficiently large absolute value of spontaneous curvature is a linactant, regardless of it partitioning between the liquid ordered and liquid disordered phase. This notion is based on the fine structure of the domain boundary revealed by a number of MD simulations (Perlmutter and Sachs, 2011; Pantano et al., 2011; Risselada and Marrink, 2008) and corroborated by the calculations in the framework of lipid membrane continuum elasticity theory (Galimzyanov et al., 2015; Staneva et al., 2016). The aforementioned ~2-4 nm shift of the boundaries of monolayer liquid-ordered domains relative to each other minimizing the deformation energy (Galimzyanov et al., 2015; Staneva et al., 2016) creates the narrow transition zone with abruptly changing geometric curvature of the surface (see Figure 6). This zone concentrates the components with both positive and negative spontaneous curvature, which drastically decreases the line tension of the boundary due to relaxation of the elastic stress and hence promotes raft fission into larger number of smaller rafts. The domain boundaries, however, can accommodate only a small amount of linactants, further increase of content of which causes its accumulation in one of the phases (or indiscriminately in both phases). This changes spontaneous curvature of the phase, which can also affect the line tension of the phase separation boundary.

Cellular membranes contain large amounts of cholesterol (up to ~40%), having large negative spontaneous curvature and partitioning almost equally between the L_o and L_d phases. Depending on the cholesterol content in the membrane and linactant properties (preferred phase, absolute value and sign of the spontaneous curvature), further increase of the linactant concentration above saturation of the boundaries can both decrease and increase the boundary line tension. E.g., the experiments in (Bao et al., 2011) were performed with the membrane containing relatively small amount of cholesterol, and GM1 ganglioside with a large positive spontaneous curvature was found to monotonously decrease the average size of domains, and hence the line tension, up to the concentrations of 5 mol%. The membranes used in the work (Galimzyanov et al., 2017b) had high cholesterol content, and in these conditions, the line tension of the boundary was found to drop abruptly up to the GM1 concentration of 0.5 mol%, where after it started to grow monotonously.

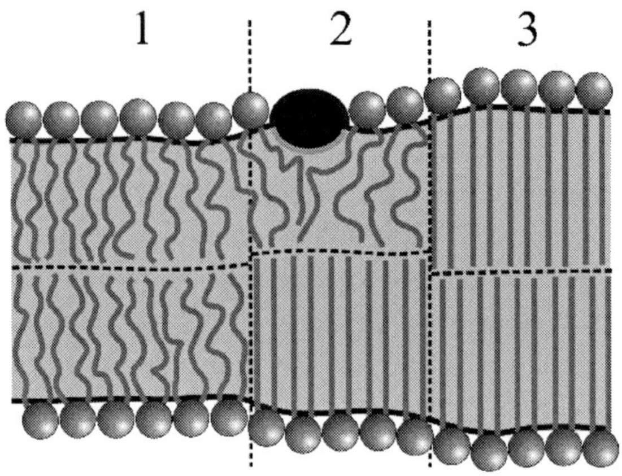

Figure 6. Preferential partitioning of membrane components. The membrane components are shown in black, including amphipatic peptides with non-zero spontaneous curvature partition into a narrow transition zone between the boundaries of liquid-ordered monolayer domains in the opposite leaflets of the membrane. 1 — Liquid-disordered phase bilayer; 2 — Transition zone (the upper monolayer is disordered, the lower is ordered); 3 — Liquid-ordered domain bilayer.

The line activity of GM1 and other gangliosides has significant physiological manifestations. Thus, the gangliosides shed by rapidly growing tumor cells are implicated in immune escape of the tumor cells and tumor overgrowth. Exogenous gangliosides were demonstrated to be able to cause raft-dependent apoptosis of the immune system cells (Molotkovskaya et al., 2002; Ladisch et al., 1987). Besides that, the non-monotonous dependence of raft boundary line tension on ganglioside concentrations appears to underlie the non-monotonous influence of GM1 on clustering of β-amyloid (Cebecauer et al., 2017).

11. Protein-Lipid Interactions in Viral Fusion and Fission

The protein-lipid interactions regulating topological rearrangements of cellular membranes, formation of vesicles, their budding and transport require coordination of a broad spectrum of proteins and associated lipids, complicating their investigation and experimental modeling. (McMahon and Gallop, 2005; Zimmerberg and Kozlov, 2006; Schekman and Orci,

1996). However, in some cases, such as the process of viral fusion, the total number of participants can be quite limited. Fusion of enveloped viruses with the target cell membrane provides a prototipical example, when membrane rearrangement and budding of the virions from the host cell membrane is achieved with the aid of a single or very few proteins (Bose et al., 2015; White and Whittaker, 2016). The enveloped viruses, represented by such diverse and formidable pathogenes as HIV, B and C hepatitus, Ebola and influenza, share certain important commonalities. All of them are wrapped in a sophisticated envelope protecting the genetic material of the virus and consisting of two layers: an outer lipid membrane (also containing integral virus proteins) and an inner protein layer. The lipids for the envelope are inherited from the host cell membrane (plasma membrane, or one of the inner membranes), whereas the inner layer is a scaffold of matrix proteins interacting both with the lipid bilayer and with the nucleocapsid of the virus. These proteins, constituting a major part of the total protein content of any enveloped virus, play an important role both in the release of the genetic material of a virus inside the host cell, and in production of the secondary virions in the infected cell (Calder et al., 2010; Ge et al., 2010; Liljeroos et al., 2011). The protein matrix serves several purposes that appear difficult to reconcile: protecting the integrity of the virus, enabling controlled release of the genetic cargo once in the host cell, and guiding assembly of new virions for the multiplication of infection. In all these processes, matrix proteins interact with lipid membranes: the outer envelope of the virus, plasma membrane of the cell or endosomal membrane. The molecular mechanisms of these interactions vital in the life cycle of the virus are far from being fully understood, the most contentious issues being related to the role of the lipid-protein interactions in the self-assembly of the viral envelope (Baudin et al., 2001; Leser and Lamb, 2017; Ruigrok et al., 2000a, 2000b; Wijesinghe et al., 2017).

Despite the large variety of shapes of the enveloped viruses (spherical, filamentous, rod-like, bullet-shaped, etc.), the spatial organization of matrix proteins in them is a lot less variable: the choice is between helical structures, as is the case with the viruses of influenza A (Calder, et al., 2010, Shtykova et al., 2013), vesicular stomatitis (Ge et al., 2010) or measles (Liljeroos et al., 2011), or two-dimensional arrays forming particles of nearly spherical shape in the case of Newcastle disease (Battisti et al., 2012) or HIV (Briggs et al., 2004). The evolutionary and structural similarities exist between the M proteins of distant viral families, e.g., between the matrix and capsid protein of HIV and M1 protein of influenza virus (Harris et al., 1999). Such a structural similarity implies commonality of the mechanisms of participation of matrix proteins in formation of the viral envelope and budding of the new virions from the host cell surface.

The virus budding process is equivalent to cellular exocytosis in the sense that a loaded vesicle is formed and detached from the plasma membrane surface. The minimalistic assortment of proteins in the viral envelope leaves the virus with a scant choice of options for affecting the membrane in order to induce its topological rearrangement. Three potential strategies can be distinguished, and are used independently or concurrently by different viruses, as schematically represented on Figure 7. The first mechanism exploits the ability of matrix proteins to induce lipid bilayer curvature upon binding it and forming a two-dimensional array on its surface due to specifics of the protein's molecular geometry. The induced curvature initiates virion formation. X-ray spectroscopy revealed this mechanism for the Newcastle disease virus (Battisti et al., 2012) and infectious salmon anaemia virus (Zhang et al., 2017). In both cases, the protein monomers formed the array at an angle corresponding to formation of nearly-spherical shape virus particles.

Formation of these particles starts with nucleation of circular proteolipid domains (Shnyrova et al., 2007) with subsequent bulging and budding of the lipid membrane. The liquid state of the lipid bilayer implies that in order to render its curvature to the lipid membrane, the matrix protein layer has to be bound to it so as to minimize redistribution of lipid at the sites of contact of the membrane with the matrix. Apparently, it has to be achieved *via* hydrophobic interactions, as was positively demonstrated in the case of the Newcastle disease virus (Faaberg and Peeples, 1988). Such an interaction mechanism was hypothesized in the earlier works dedicated to investigation of influenza A M1 protein interactions with the lipid bilayer (Gregoriades, 1980; Gregoriades and Frangione, 1981; Ye et al., 1987; Sha and Luo, 1997), however it was not corroborated by later structural investigations (Baudin et al., 2001, Shtykova et al., 2013), which provided no evidence for partial incorporation of the M1 protein into the lipid bilayer, nor for its ability to dictate a certain curvature to the membrane. These latter studies provide arguments in favor of the electrostatic nature of interaction of the M1 protein with the membrane.

The second possible mechanism is condensation of lipid under the protein layer induced by electrostatic interactions. The condensation on the inner monolayer of the plasma membrane would create a local imbalance of the area of membrane leaflets resulting in membrane bending and formation of various protrusions serving to relief the excess surface (Shi and Baumgart, 2015). In particular, formation of filamentous structures with the diameter comparable to that of virus particles upon adsorption of influenza C virus matrix protein was observed in experiments with negatively charged giant unilamellar vesicles (GUV) (Saletti et al., 2017). Electrostatic interactions with anionic lipids is inherent to matrix proteins of influenza (Batishchev et al., 2016; Baudin et al., 2001; Hilsch et al., 2014; Shilova et al., 2017), Ebola (Adu-Gyamfi et al., 2015; Gerstman et al., 2016), and vesicular stomatitis (Justice et al., 1995) viruses. Matrix proteins of all these viruses was demonstrated to be capable of forming virus-like particles (Gómez-Puertas et al., 2000; Latham and Galarza, 2001; Jasenosky et al., 2001; Justice et al., 1995). Any additional factors in favor of change in the membrane curvature would facilitate virus particle budding per this mechanism. Thus, palmitoylation of influenza A virus hemagglutinin (HA) facilitating creation of the necessary curvature of the viral particle was demonstrated to enhance M1 protein adsorption on the lipid membrane and facilitate the process of formation of virus-like particles (Chlanda et al., 2017).

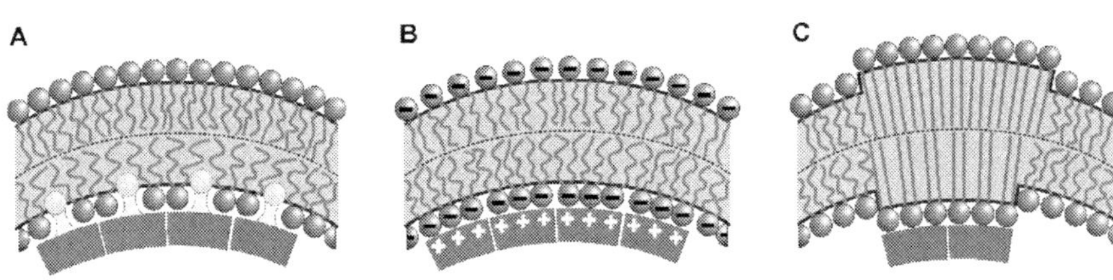

Figure 7. Mechanisms of induction of membrane curvature by matrix proteins during formation of virions in the infected cell. **A**, At the expense of partial incorporation into the membrane imposing the molecular geometry of a 2D array of matrix proteins. **B**, Condensation of lipids in one of the membrane monolayers at the expense of electrostatic interactions between matrix protein and lipid molecules. **C**, Formation of a liquid-ordered lipid domain caused by interaction with matrix proteins and subsequent budding induced by line tension of the domain boundary.

Finally, the lipid membranes of a number of enveloped viruses have been demonstrated to be more detergent-resistant than the plasma membranes of their host cells (Veit and Thaa B, 2011; Scheiffele et al., 1999; Gerl et al., 2012; Nayak et al., 2009; Pohl et al., 2007), suggesting that the viral envelopes might be budding from the raft phase. From this, another mechanism of viral particle vesiculation and budding can be inferred based on the increase of lateral tension of the raft boundary causing the raft to bulge and protrude from the plane of the membrane (Baumgart et al., 2003). Although to date such a process has not been experimentally reported for any enveloped virus, the lipid domains with the required properties can be formed *via* condensation of negatively charged lipids with fully saturated acyl chains on the inner leaflet of the cellular plasma membrane with concomitant initiation of formation of sphingomyelin-enriched lipid domain on the outer leaflet (Bobone et al., 2017).

12. Curvature-Coupled Membrane-Protein Interactions

The membrane curvature stands out from the rest of the discussed parameters describing the state of the lipid membrane as more of a geometric than a physical property. However, the existence of a variety of proteins either sensing or regulating membrane curvature assigns cellular membrane shape/geometrical curvature to act as a complementary modus operandi for regulating membrane processes and functions (Madsen et al., 2010; Baumgart et al., 2011; Bigay & Antonny, 2012; McMahon & Boucrot, 2015; Iversen et al., 2015). Any type of hydrophobic moiety of a protein embedded into lipid bilayer can provide coupling between the membrane shape and the protein function (Bhatia et al., 2010). Change of the elastic stress/lateral pressure profile induced by the membrane curvature can affect these hydrophobic anchors, inducing conformational transitions (Figure 8A) and changes of intrinsic activity, as it was established, e.g., for cytidine triphosphate: phosphocholine cytidylyl-transferase (CCT), an enzyme catalyzing the rate limiting reaction in the phosphatidylcholine biosynthesis (Saita & Mendoza, 2015). In the soluble form, the enzyme is autoinhibited by an amphipathic helix, capable of binding the bilayers with a high content of lipids inducing negative spontaneous curvature (such as phosphoethanolamine (PE)). Interaction with the membrane relieves the autoinhibition and drastically enhances PC synthesis, making the CCT a perfect negative feedback protection system, adjusting the PC biosynthesis to the demand to allow the membrane expansion or prevent inadvertent pore formation (Attard et al., 2000). Such a scenario implies that the protein has at least two conformations, sensitivity of which to the membrane curvature exceeds the conformational energy difference. If this option is not available, and there is no conformational transition that can relief the energy frustration caused by curvature, the protein can simply redistribute between the more curved and the less curved regions (Figure 8B). Besides that, appearance of lipid packing defects in the regions of positive curvature can reduce the energy barrier for protein insertion into the lipid bilayer (Vanni et al., 2014). Thus, membrane curvature has a potential for regulating both the activity and the spatial organization of membrane proteins.

Membrane recruitment of lipidated proteins, such as Ras GTPases, is regulated by the synergy of the membrane curvature and lipid composition (Larsen et al., 2017): partitioning of the protein into the curved membrane areas is stronger where bending leads to a more pronounced change of the lateral pressure profile. There are many endocytic peripheral proteins, e.g., ENTH (Capraro et al., 2010), endophilin (Simunovic et al., 2016), dynamin (Shnyrova et al., 2013), shallow insertions of amphiphatic helices of which into the lipid bilayer force them to accumulate in convex regions of membranes. Such shallow-embedding curvature sensors can also act as curvature generators at high bulk concentrations (Bhatia et al., 2010). The ability of some endocytic proteins to generate membrane curvature is thought to be crucial in membrane remodeling during endo/exocytosis (Kozlov et al., 2010). Their accumulation on the surface of the membrane of a growing vesicle/initial invagination should facilitate membrane budding (Figure 8C). The subsequent fission of the membrane neck connecting the vesicle and the parent membrane requires generation of an extremely curved precursor state, the luminal radius of which is comparable with the monolayer thickness (Bashkirov et al., 2008). For such extreme curvatures to be attained, an additional source of force squeezing the membrane neck is required.

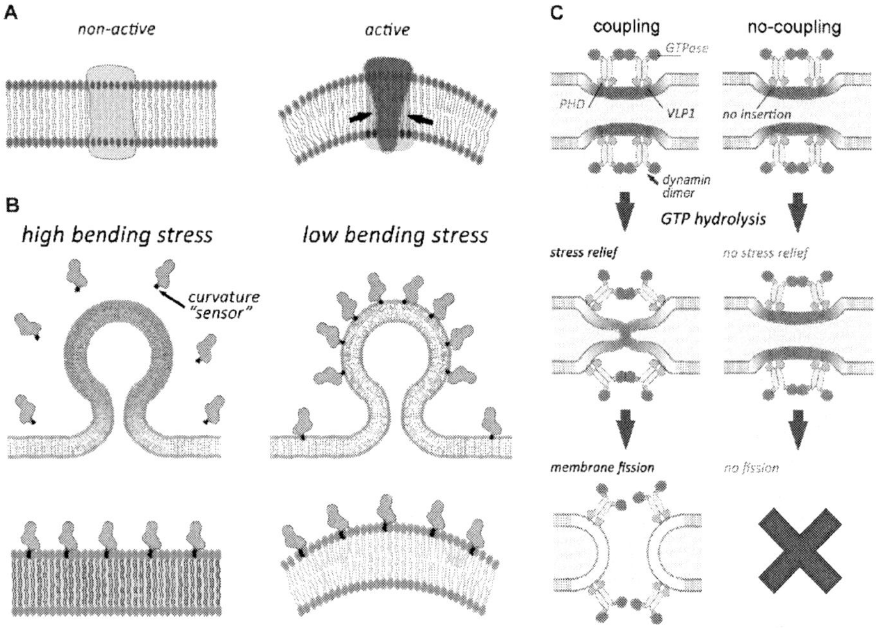

Figure 8. Curvature-coupled membrane-protein interactions. **A**, Change of lateral pressure profile induces conformational transition of a transmembrane protein. **B**, Any hydrophobic moiety a protein inserts into a lipid bilayer works as a “sensor” of the membrane bending stress capable of driving proteins to redistribute in the curvature gradient. Reduction of the lipid bilayer elastic stress related to proteins accumulation in the curved region of the membrane counterpoises their entropy-driven scattering. Inversely, the elastic stress induced in a flat lipid bilayer by insertion of amphipathic helices into it can cause membrane bending, leading to its stress relief. **C**, Membrane inserting variable loop 1 (VLP1) of the pleckstrin homology domain (PHD) of dynamin provides a strong coupling between the membrane curvature and dynamin, which is crucial for the membrane fission catalysis by the protein. GTP hydrolysis induced tilting of the PHDs of dynamin allowing it to conform with the membrane deformations creates a low-energy pathway for the membrane fission. Reduction of curvature coupled dynamin-membrane interaction inhibits the membrane fission catalyzed by dynamin, without altering its ability to polymerize and to squeeze the membrane neck.

One of such sources is the entropic force originating from the collisions between proteins in densely populated (“crowded”) membranes. High surface concentration of proteins could create lateral pressure in the plane of protein-protein interactions, the surface of which protrudes from the membrane. That in turn results in a pronounced increase of the membrane lateral tension (the so-called crowding effect) (Stachowiak et al., 2010). Thus, crowding explains the membrane fission caused *in vitro* by the proteins, the known functions of which do not include membrane fission (Snead et al., 2017). Moreover, crowding does not generally require shallow insertion, which merely facilitates the protein recruitment on the membrane (Snead et al., 2017). A much more elegant membrane fission mechanism of GTPases from dynamin family is based on the coupling between membrane bending stress relief and orientation of the curvature-inducing motifs of the protein (Schmid & Frolov, 2011). Historically, the mechano-chemical activity of Dynamin1 involved in the membrane fission associated with synaptic recycling is better studied (Antonny et al., 2016).

The results of investigation of the fission of lipid nanotubes (NT) by dynamin1 *in vitro* allows dividing the protein action into two distinct steps. The first of them is squeezing of the target NT by protein polymerization into contractile rings closely adjacent to each other on the surface of the membrane neck (this step does not require GTP hydrolysis) (Hinshaw & Schmid, 1995). The second step is the subsequent catalysis of the membrane fission reaction, which requires cooperative GTP hydrolysis activated by dynamin GTPase domains dimerization between the adjacent rings (Chappie et al., 2010). The mechanical stress pumped into lipid bilayer at the first stage is sufficient to approach the NT radii at which spontaneous membrane fission could occur, but the dynamin scaffold fixing the membrane geometry prevents the membrane from remodeling (Bashkirov et al., 2008). Cooperative GTP hydrolysis induces disassembly of the dynamin structure and the membrane fission in a stochastic manner (Shnyrova et al., 2013). Membrane fission becomes possible due to catalytic role of the dyanmin1 membrane-interacting pleckstrin homology domain (PHD) (Shnyrova et al., 2013), variable loop 1 (VL1) of which shallowly penetrate into the lipid bilayer (Chappie et al., 2011). The Ile → Ala (I533A) mutation in VL1 of the PHD suppresses membrane fission without altering dynamin's polymerization or GTPase activity (Shnyrova et al., 2013), revealing significance of curvature-coupled membrane protein interaction in dynamin1 membrane fission. Shallow insertion of the PH domains into lipid bilayer forms a circle of wedges around the outer monolayer, their initial orientation, while dynamin is in the polymerized form, being normal to the neck axis (Figure 8D). After GTP hydrolysis and subsequent disassembly of the protein scaffold, these wedges acquire an additional degree of freedom in their orientation in the monolayer. The elastic stress accumulated in the curved membrane forces the PH wedges of two adjacent rings to rotate in the opposite directions leading to additional sagging of the membrane neck between them (Figure 8D). Such a transient structure has a significantly lower energy barrier for hemifission (Shnyrova et al., 2013; Mattila et al., 2015). The subsequent disassembly of the dynamin polymer leads to a complete transition from hemifission into fission due to lateral tension (Mattila et al., 2015).

Thus, dynamin acts as a classical catalyzer in the sense that it does not expend energy during the fission reaction and only spatially organizes the reaction intermediate so as to reduce its free energy in order to significantly increase the probability of transition of the fission precursor into fission. Dynamin 1 has a very high constriction ability due to its high rate of polymerization, so that at the high GTP concentrations only the structural organization (two rung dynamin scaffold) geometrically optimized for fission can assemble (Shnyrova et al., 2013). This explains the high potency of dynamin 1 in inducing the membrane fission, which appears to be vital for neurons.

13. Current Vision, Obstacles and Outlook

The structural versatility of proteins made of a limited number of different kinds of building blocks (amino acid residues) is virtually unlimited due to ability of the polymeric chain to adopt a broad spectrum of secondary, ternary and quaternary structures stabilized by diverse interactions of their backbone and side chains. However, once synthesized and folded, a given protein is typically confined to a very narrow subset of its conformational space, departure from which could be an irreversible process known as denaturation. Being substantially larger than lipids and associating rather tightly to surrounding lipid shells by virtue of specific interactions and wetting phenomena, transmembrane and peripheral proteins usually interact with a relatively small number of closest neighbors in the membrane and water-soluble counterparts.

The number of the building blocks of the lipid subsystem of the membrane – i.e., distinct lipid species, though much smaller than that of membrane proteins, significantly exceeds the number of amino acids. In addition, within fluid membrane the mobility and combinatorial interactions of lipids are only kinetically limited, and all the lipidic components of the membrane are capable of dynamically entering kaleidoscopic mutual interactions and forming temporarily immiscible localized domains of the composition substantially different from the bulk membrane. Such phase separation can be triggered and regulated by temperature, localized changes of ions concentrations, local mechanical stress or change of geometric shape of the membrane, local electrostatics, and by specific or non-specific interactions with integral or peripheral membrane proteins that can change in response to functionally relevant conformational transitions.

The lateral inhomogeneity of cellular membranes is admittedly much more complex and functionally significant than could be assumed less than two decades ago when the existence of membrane rafts was positively demonstrated for the first time. Based on all the experience accumulated to date, any definition of a "raft" falls short of providing precise and unambiguous criteria of attribution to "raft" or "non-raft" category. Nevertheless, the term is still widely (and conveniently) used, though in a way more inclusive than definitive, describing a generic phenomenon rather than a distinct object with a fully determined identity. Every raft has a face, or rather two faces, and the faces have an assortment of facial expressions, some of which are mere whims or fluctuations, while other ones are indicative of (and significant for) the function about to be performed by the raft or a subset of its constituents. The "expressions" can be characterized, and often even quantified, with a (hopefully) limited set of "mimic traits" (line tension, lateral pressure profile, order parameter, dielectric permittivity, distribution of surface charge density, presence of specifically interacting rare lipid species, local geometric curvature, spontaneous curvature of each monolayer etc.). The states of the opposing monolayers are not independent of each other, but the existing dependence is not trivial, ranging from strong attraction driving the opposing domains into complete registry (or almost complete registry with a slight shift to minimize bending energy in the transition zone) to mutually exclusive repulsion. As suggested by several examples discussed above, sometimes this dependence alone can mediate a vital biological function, and there are reasons to believe that it is widely exploited by nature for fine regulation of such functions and coordination between them.

This does not subtract from the value of the reductionist approach, since like human faces, all the rafts and their dynamically changing states have essential commonalities allowing the functions of isolated proteins to be studied in simplified *in vitro*, *in silico*, and theoretical models with manageable numbers of the controllable and minimal number of the uncontrollable variables. However, only a holistic vision of the protein-lipid ensemble behavior can capture all the subtleties of regulation of individual functions and coordination between the interdependent functions through the effects of dynamically changing and spatially heterogeneous properties of lipid environment upon the conformational dynamics of the embedded, partly submerged, and surface-bound or transiently absorbing proteins. The interactions of consequence for the functions of a protein are not limited to the lipids and other proteins in direct contact with it, but luckily, their effective spatial and temporal range is rather limited, and only a few interaction modes and physical mechanisms appear to be essential in the case of each particular protein or its biological function. Given the formidable and continuously growing arsenal of the experimental methods capable of resolving local changes of lipid environment properties (EPR, NMR, classical fluorescent and FRET-FLIM microscopy, differential scanning calorimetry, and a range of biochemical tools, such as bifunctional lipids), increasing number of solutions to pending issues and puzzles of the mechanisms of biological functions performed by membrane-associated proteins can be confidently predicted to be found in the realm of such interactions. Some of the new mechanisms enabling the orchestrated interplay between the components of biological membranes suggested in this review, such as competitive binding of specifically interacting lipids with conformation-dependent binding activity, are merely plausible hypothesis, while some other, like transmembrane signaling based on transmonolayer correlation of lipid environment local properties or multiple roles of lipid domains and linactants in biological processes, are corroborated by compelling, though circumstantial, physiological, experimental and theoretical results. Whether or not all these hypotheses will stand to an ultimate experimental test, if they inspire an experimentalist or theoretician to do better justice to the diverse biological roles of lipid-mediated dynamic processes, the authors would consider their efforts well rewarded.

ACKNOWLEDGMENTS

The authors express their sincere thanks to Dr. K.A. Beirit for helpful discussions.

Funding Sources

This work was supported by Russian Foundation for Basic Research grants #15-04-07983 (Sections 1 and 2), #18-04-01289 (Section 3), #17-04-02045 (Sections 4 and 5), #15-54-15006 (Sections 6-9), #17-04-02070 (Section 10), #17-54-30022 (Section 11), and #17-04-02042 (Section 12). O.V.B. was supported by the grant of the President of the Russian Federation MK-6058.2016.4. Support from the RAS Programme "Molecular and Cellular Biology" is greatly appreciated.

Author Contributions

The manuscript was written through author contributions as follows: Sections 1-13 by K.V.P. and E.V.B.; Sections 6-10 by S.A.A.; Section 11 by O.V.B.; Section 12 and 13 by P.V.B. and K.V.C. All authors have given approval to the final version of the manuscript.

REFERENCES

Adu-Gyamfi, E., Johnson, K. A., Fraser, M. E., Scott, J. L., Soni, S. P., Jones, K. R., Digman, M. A., Gratton, E., Tessier, C. R., Stahelin, R. V., 2015. Host cell plasma membrane phosphatidylserine regulates the assembly and budding of Ebola virus. *J. Virol.* 89, 9440-9453.

Aguilar, P. S., Cronan, J. E., de Mendoza, D., 1998. A Bacillus subtilis gene induced by cold shock encodes a membrane phospholipid desaturase. *J. Bacteriol.* 180, 2194–2200.

Akimov S. A., Frolov V. A., Kuzmin P. I., Zimmerberg J., Chizmadzhev Yu. A., Cohen F. S. 2008. Domain formation in membranes caused by lipid wetting of protein. *Phys. Rev. E* 77, 051901.

Anderson, R. G., Jacobson, K., 2002. A role for lipid shells in targeting proteins to caveolae, rafts, and other lipid domains. *Science* 296, 1821-1825.

Antonny, B., Burd, C., De Camilli, P., Chen, E., Daumke, O., Faelber, K., Ford M., Frolov V. A., Frost A., Hinshaw J. E., Kirchhausen T., Kozlov M. M., Lenz M., Low H. H., McMahon H., Merrifield C., Pollard T. D., Robinson P. J., Roux A., Schmid S., 2016. Membrane fission by dynamin: what we know and what we need to know. *EMBO J.* 35, 2270-2284.

Attard, G. S., Templer, R. H., Smith, W. S., Hunt, A. N., Jackowski, S., 2000. Modulation of CTP: phosphocholine cytidylyltransferase by membrane curvature elastic stress. *Proc. Natl. Acad. Sci. U. S. A.* 97, 9032–9036.

Ayuyan, A. G., Cohen, F. S., 2006. Lipid peroxides promote large rafts: effects of excitation of probes in fluorescence microscopy and electrochemical reactions during vesicle formation. *Biophys. J.* 91, 2172-2183.

Ayuyan, A. G., Cohen, F. S., 2008. Raft composition at physiological temperature and pH in the absence of detergents. *Biophys. J.* 94, 2654-2666.

Bao, R., Li, L., Qiu, F., Yang, Y., 2011. Atomic force microscopy study of ganglioside GM1 concentration effect on lateral phase separation of sphingomyelin/dioleoylphosphatidylcholine/cholesterol bilayers. *J. Phys. Chem. B* 115, 5923-5929.

Barrett, P. J., Song Y., Van Horn, W. D., Hustedt, E. J., Schafer, J. M., Hadziselimovic, A., Beel, A. J., Sanders, C. R. The amyloid precursor protein has a flexible transmembrane domain and binds cholesterol. *Science* 2012 336, 1168-1171.

Barwe, S. P., Anilkumar, G., Moon, S. Y., Zheng, Y., Whitelegge, J. P., Rajasekaran, S.A., Rajasekaran, A.K., 2005. Novel role for Na, K-ATPase in phosphatidylinositol 3-kinase signaling and suppression of cell motility. *Mol. Biol. Cell* 16, 1082-1094.

Bashkirov, P. V., Akimov, S. A., Evseev, A. I., Schmid, S. L., Zimmerberg, J., Frolov, V. A., 2008. A partnership between dynamin and lipids defines dynamics and intermediates of membrane fission. *Cell* 135, 1276-1286.

Batishchev, O. V., Shilova, L. A., Kachala, M. V., Tashkin, V. Y., Sokolov, V. S., Fedorova, N. V., Knyazev, D. G., Zimmerberg, J., Chizmadzhev, Y. A., 2016. pH-dependent formation and disintegration of the influenza A virus protein scaffold to provide tension for membrane fusion. *J. Virol.* 90, 575-585.

Battisti, A. J., Meng, G., Winkler, D. C., McGinnes, L. W., Plevka, P., Steven, A. C., Morrison, T.G., Rossmann, M.G., 2012. Structure and assembly of a paramyxovirus matrix protein. *Proc. Natl. Acad. Sci. U. S. A.* 109, 13996-14000.

Baudin, F., Petit, I., Weissenhorn, W., Ruigrok, R. W. H., 2001. In vitro dissection of the membrane and RNP binding activities of influenza virus M1 protein. *Virology* 281, 102-108.

Baumgart, T., Capraro, B. R., Zhu, C., Das, S. L., 2011. Thermodynamics and mechanics of membrane curvature generation and sensing by proteins and lipids. *Annu. Rev. Phys. Chem.* 62, 483-506.

Baumgart, T., Hess, S. T., Webb, W. W., 2003. Imaging coexisting fluid domains in biomembrane models coupling curvature and line tension. *Nature* 425, 821-824.

Bhatia, V. K., Hatzakis, N. S., Stamou, D., 2010. A unifying mechanism accounts for sensing of membrane curvature by BAR domains, amphipathic helices and membrane-anchored proteins. *Semin. Cell Dev. Biol.* 21, 381-390.

Bianconi, S. E., Cross, J. L., Wassif, C. A., Porter, F. D., 2015. Pathogenesis, epidemiology, diagnosis and clinical aspects of Smith-Lemli-Opitz syndrome. *Expert Opin. Orphan Drugs* 3, 267–280.

Bigay, J., Antonny, B., 2012. Curvature, lipid packing, and electrostatics of membrane organelles: defining cellular territories in determining specificity. *Dev. Cell* 23, 886-895.

Bobone, S., Hilsch, M., Storm, J., Dunsing, V., Herrmann, A., Chiantia, S., 2017. Phosphatidylserine lateral organization influences the interaction of influenza virus matrix protein 1 with lipid membranes. *J. Virol.* 91, e00267-17.

Bocharov, E. V., Bragin, P. E., Pavlov, K. V., Bocharova, O. V., Mineev, K. S., Polyansky, A. A., Volynsky, P. E., Efremov, R. G., Arseniev, A. S., 2017a. The conformation of the Epidermal Growth Factor Receptor transmembrane domain dimer dynamically adapts to the local membrane environment. *Biochemistry* 56, 1697-1705.

Bocharov, E. V., Korzhnev, D. M., Blommers, M. J. J., Arvinte, T., Orekhov, V. Yu, Billeter, M, Arseniev, A. S., 2002. Dynamics-modulated biological activity of transforming growth factor β3. *J. Biol. Chem.* 277, 46273-46279.

Bocharov, E. V., Lesovoy, D. M., Goncharuk, S. A., Goncharuk, M. V., Hristova, K., Arseniev, A. S., 2013. Structure of FGFR3 transmembrane domain dimer: implications for signaling and human pathologies. *Structure* 21, 2087-2093.

Bocharov, E. V., Lesovoy, D. M., Pavlov, K. V., Pustovalova, Y. E., Bocharova, O. V., Arseniev, A. S., 2016. Alternative packing of EGFR transmembrane domain suggests that protein-lipid interactions underlie signal conduction across membrane. *Biochim. Biophys. Acta* 1858, 1254-1261.

Bocharov, E. V., Mayzel, M. L., Mineev, K. S., Tkach, E. N., Ermolyuk, Ya. S., Schulga, A. A., Arseniev, A. S., 2010b. Left-handed dimer of EphA2 transmembrane domain: helix packing diversity among receptor tyrosine kinases. *Biophys. J.* 98, 881-889.

Bocharov, E. V., Mineev, K. S., Pavlov, K. V., Akimov, S. A., Kuznetsov, A. S., Efremov, R. G., Arseniev, A. S., 2017b. Helix-helix interactions in membrane domains of bitopic proteins: specificity and role of lipid environment. *Biochim. Biophys. Acta* 1859(9 Pt A), 1417-1429.

Bocharov, E. V., Pavlov, K. V., Blommers, M. J. J., Arvinte, T., Arseniev, A. S., 2008. Modulation of the Bioactive Conformation of Transforming Growth Factor beta: Possible Implications of Cation Binding for Biological Function. *Bioactive Conformation II, Series: Topics in Current Chemistry*, 273, 155-182.

Bocharov, E. V., Sharonov, G. V., Bocharova, O. V., Pavlov, K. V., 2017c. Conformational transitions and interactions underlying the function of membrane embedded receptor protein kinases. *Biochim. Biophys. Acta.* 1859(9 Pt A):1417-1429.

Bocharov, E. V., Volynsky, P. E., Pavlov, K. V., Efremov, R. G., Arseniev, A. S., 2010a. Structure elucidation of dimeric transmembrane domains of bitopic proteins. *Cell Adh. Migr.* 4, 284-298.

Bose, S., Jardetzky, T. S., Lamb, R. A., 2015. Timing is everything: Fine-tuned molecular machines orchestrate paramyxovirus entry. *Virology* 479, 518-531.

Bragin, P. E., Mineev, K. S., Bocharova, O. V., Volynsky, P. E., Bocharov E. V., Arseniev, A. S., 2016. HER2 transmembrane domain dimerization coupled with self-association of membrane-embedded cytoplasmic juxtamembrane regions. *J. Mol. Biol.* 428, 52-61.

Bretscher, M. S., Munro, S., 1993. Cholesterol and the Golgi apparatus. *Science* 261, 1280–1281.

Brewster, R., Safran, S. A., 2010. Line active hybrid lipids determine domain size in phase separation of saturated and unsaturated lipids. *Biophys. J.* 98, L21-L23.

Briggs, J. A., Simon, M. N., Gross, I., Krausslich, H. G., Fuller, S. D., Vogt, V. M., Johnson, M. C., 2004. The stoichiometry of Gag protein in HIV-1. *Nat. Struct. Mol. Biol.* 11, 672-676.

Bugge, K., Lindorff-Larsen, K., Kragelund, B. B., 2016. Understanding single-pass transmembrane receptor signaling from a structural viewpoint-what are we missing? *FEBS J.* 283, 4424-4451.

Bustillo-Zabalbeitia, I., Montessuit, S., Raemy, E., Basañez, G., Terrones, O., Martinou, J. C. 2014. Specific interaction with cardiolipin triggers functional activation of Dynamin-Related Protein 1. *PLoS One.* 9, e102738.

Butters, T. D., 2007. Gaucher disease. *Curr. Opin. Chem. Biol.* 11, 412-418.

Calder, L. J., Wasilewski, S., Berriman, J. A., Rosenthal, P. B., 2010. Structural organization of a filamentous influenza A virus. *Proc. Natl. Acad. Sci. U. S. A.* 107, 10685-10690.

Capraro, B. R., Yoon, Y., Cho, W., Baumgart, T., 2010. Curvature sensing by the epsin N-terminal homology domain measured on cylindrical lipid membrane tethers. *J. Am. Chem. Soc.* 132, 1200-1201.

Cebecauer, M., Hof, M., Amaro, M., 2017. Impact of GM 1 on membrane-mediated aggregation/oligomerization of β-amyloid: Unifying view. *Biophys. J.* 113, 1-6.

Chaplin, R., Thach, L., Hollenberg, M. D., Cao, Y., Little, P. J., Kamato, D., 2017. Insights into cellular signalling by G protein coupled receptor transactivation of cell surface protein kinase receptors. *J. Cell Commun. Signal.* 11, 117-125.

Chappie, J. S., Acharya, S., Leonard, M., Schmid, S. L., Dyda, F., 2010. G domain dimerization controls dynamin's assembly-stimulated GTPase activity. *Nature* 465, 435-440.

Chlanda, P., Mekhedov, E., Waters, H., Sodt, A., Schwartz, C., Nair, V., Blank, P. S., Zimmerberg, J., 2017. Palmitoylation contributes to membrane curvature in Influenza A virus assembly and hemagglutinin-mediated membrane fusion. *J. Virol.* 91, e00947-17.

Cornelius, F., 2001. Modulation of Na,K-ATPase and Na-ATPase Activity by Phospholipids and Cholesterol. I. Steady-State Kinetics. *Biochemistry* 40, 8842-8851.

Cybulski, L. E., Martín, M., Mansilla, M. C., Fernández, A., de Mendoza. D., 2010. Membrane thickness cue for cold sensing in a bacterium. *Curr. Biol.* 20, 1539-1544.

Dan, P. D., Thuong, T. T., Minh, P. D., Loi, D. D., Nhi, N. B., Van Chi, P., 2013. Analysis of the Membrane Proteins in Human Serum. *J. Proteomics Bioinform.* 6, 296-301.

De Mendoza, D., 2014. Temperature sensing by membranes. *Annu. Rev. Microbiol.* 68, 101–116.

DeBarber, A. E., Eroglu, Y., Merkens, L. S., Pappu, A. S., Steiner, R. D., 2011. Smith-Lemli-Opitz syndrome. *Expert Rev. Mol. Med.* 13, e24.

Doerner, A., Scheck, R., Schepartz, A., 2015. Growth factor identity is encoded by discrete coiled-coil rotamers in the EGFR juxtamembrane region. *Chem. Biol.* 22, 776-784.

Edidin M., 2001. Shrinking patches and slippery rafts: scales of domains in the plasma membrane. *Trends Cell Biol.* 11, 492-496.

Endres, N. F., Das, R., Smith, A. W., Arkhipov, A., Kovacs, E., Huang, Y., Pelton, J. G., Shan, Y., Shaw, D. E., Wemmer, D. E., Groves, J. T., Kuriyan J., 2013. Conformational coupling across the plasma membrane in activation of the EGF receptor. *Cell* 152, 543–556.

Epand, R. M., Sayer, B. G., Epand, R. F., 2005. Caveolin scaffolding region and cholesterol-rich domains in membranes. *J. Mol. Biol.* 345, 339-350.

Faaberg, K. S., Peeples, M. E., 1988. Association of soluble matrix protein of Newcastle disease virus with liposomes is independent of ionic conditions. *Virology* 166, 123–132.

Falkenburger, B. H., Jensen, J. B., Dickson, E. J., Suh, B. C., Hille, B., 2010. Phosphoinositides: lipid regulators of membrane proteins. *J. Physiol.* 588(Pt 17), 3179-3185.

Feigenson, G. W., Buboltz, J. T., 2001. Ternary phase diagram of dipalmitoyl-PC/dilauroyl-PC/cholesterol: nanoscopic domain formation driven by cholesterol. *Biophys. J.* 80, 2775-2788.

Fowler, P. W., Williamson, J. J., Sansom, M. S., Olmsted, P. D., 2016. Roles of interleaflet coupling and hydrophobic mismatch in lipid membrane phase-separation kinetics. *J. Am. Chem. Soc.* 138, 11633-11642.

Frolov, V. A. J., Chizmadzhev, Y. A., Cohen, F. S., Zimmerberg, J., 2006. "Entropic traps" in the kinetics of phase separation in multicomponent membranes stabilize nanodomains. *Biophys. J.* 91, 189-205.

Galimzyanov, T. R., Kuzmin, P. I., Pohl, P., Akimov, S. A., 2017a. Undulations drive domain registration from the two membrane leaflets. *Biophys. J.* 112, 339-345.

Galimzyanov, T. R., Lyushnyak, A. S., Aleksandrova, V. V., Shilova, L. A., Mikhalyov, I. I., Molotkovskaya, I. M., Akimov, S. A., Batishchev, O.V., 2017b. Line activity of ganglioside GM1 regulates the raft size distribution in a cholesterol-dependent manner. *Langmuir* 33, 3517-3524.

Galimzyanov, T. R., Molotkovsky, R. J., Bozdaganyan, M. E., Cohen, F. S., Pohl, P., Akimov, S. A., 2015. Elastic membrane deformations govern interleaflet coupling of lipid-ordered domains. *Phys. Rev. Lett.* 115, 088101.

García-Marcos, M., Pochet, S., Tandel, S., Fontanils, U., Astigarraga, E., Fernández-González, J. A., Kumpsd, A., Aida Marinoa, A., Dehaye, J. P., 2006. Characterization and comparison of raft-like membranes isolated by two different methods from rat submandibular gland cells. *Biochim. Biophys. Acta Biomembr.* 1758, 796-806.

Ge, P., Tsao, J., Schein, S., Green, T. J., Luo, M., Zhou, Z. H., 2010. Cryo-EM model of the bullet-shaped vesicular stomatitis virus. *Science* 327, 689-693.

Gerl, M. J., Sampaio, J. L., Urban, S., Kalvodova, L., Verbavatz, J. M., Binnington, B., Lindemann, D., Lingwood, C. A., Shevchenko, A., Schroeder, C., Simons, K., 2012. Quantitative analysis of the lipidomes of the influenza virus envelope and MDCK cell apical membrane. *J. Cell. Biol.* 196, 213-221.

Gerstman, B. S., Stahelin, R. V., Chapagain, P. P., 2016. The Ebola virus protein VP40 hexamer enhances the clustering of PI (4, 5) P 2 lipids in the plasma membrane. *Phys. Chem. Chem. Phys.* 18, 28409-28417.

Gil, T., Ipsen, J. H., 1997. Capillary condensation between disks in two dimensions. *Phys. Rev. E* 55, 1713-1721.

Gil, T., Mikheev, L. V., 1995. Curvature controlled wetting in two dimensions. *Phys. Rev. E* 52, 772-780.

Gil, T., Sabra, M. C., Ipsen, J. H., Mouritsen, O. G., 1997. Wetting and capillary condensation as means of protein organization in membranes. *Biophys. J.* 73, 1728-1741.

Gómez, J., Sagués, F., Reigada, R., 2009. Nonequilibrium patterns in phase-separating ternary membranes. *Phys. Rev. E* 80, 011920.

Gómez-Puertas, P., Albo, C., Pérez-Pastrana, E., Vivo, A., Portela, A., 2000. Influenza virus matrix protein is the major driving force in virus budding. *J. Virol.* 74, 11538-11547.

Grauffel, C., Yang, B., He, T., Roberts, M. F., Gershenson, A., Reuter, N., 2013. Cation-π interactions as lipid-specific anchors for phosphatidylinositol-specific phospholipase C. *J. Am. Chem. Soc.* 135, 5740-5750.

Gregoriades, A., 1980. Interaction of influenza M protein with viral lipid and phosphatidylcholine vesicles. *J. Virol.* 36, 470-479.

Gregoriades, A., Frangione, B., 1981. Insertion of influenza M protein into the viral lipid bilayer and localization of site of insertion. *J. Virol.* 40, 323-328.

Haataja, M. P., 2017. Lipid domain co-localization induced by membrane undulations. *Biophys. J.* 112, 655-662.

Harris, A., Sha, B., Luo, M., 1999. Structural similarities between influenza virus matrix protein M1 and human immunodeficiency virus matrix and capsid proteins: an evolutionary link between negative-stranded RNA viruses and retroviruses. *J. Gen. Virol.* 80, 863-869.

Harris, J. S., Epps, D. E., Davio, S. R., Kezdy, F. J., 1995. Evidence for transbilayer, tail-to-tail cholesterol dimers in dipalmitoylglycerophosphocholine liposomes. *Biochemistry* 34, 3851-3857.

He, C., Hobert, M., Friend, L., Carlin, C., 2002. The epidermal growth factor receptor juxtamembrane domain has multiple basolateral plasma membrane localization determinants, including a dominant signal with a polyproline core. *J. Biol. Chem.* 277, 38284-38293.

Heberle, F. A., Marquardt, D., Doktorova, M., Geier, B., Standaert, R. F., Heftberger, P., Kollmitzer, B., Nickels, J. D., Dick, R. A., Feigenson, G. W., Katsaras, J., London, E., Pabst, G., 2016. Subnanometer structure of an asymmetric model membrane: interleaflet coupling influences domain properties. *Langmuir* 32, 5195-5200.

Heo, W. D., Inoue, T., Park, W. S., Kim, M. L., Park, B. O., Wandless, T. J., Meyer, T., 2006. PI(3,4,5)P3 and PI(4,5)P2 lipids target proteins with polybasic clusters to the plasma membrane. *Science* 314, 1458–1461.

Hilgemann, D. W., Ball. R., 1996. Regulation of cardiac Na+,Ca2+ exchange and KATP potassium channels by PIP2. *Science* 273, 956-959.

Hilsch, M., Goldenbogen, B., Sieben, C., Höfer, C. T., Rabe, J. P., Klipp, E., Herrmann, A., Chiantia, S., 2014. Influenza A matrix protein M1 multimerizes upon binding to lipid membranes. *Biophys. J.* 107, 912-923.

Hinshaw, J. E., Schmid, S. L., 1995. Dynamin self-assembles into rings suggesting a mechanism for coated vesicle budding. *Nature* 374, 190-192.

Hirama, T., Das, R., Yang, Y., Ferguson, C., Won, A., Yip, C. M., Kay, J. G., Grinstein, S., Parton, R. G., Fairn, G.D., 2017. Phosphatidylserine dictates the assembly and dynamics of caveolae in the plasma membrane. *J. Biol. Chem.* 292, 14292-14307.

Holmgren, J., Lönnroth, I., Månsson, J., Svennerholm, L., 1975. Interaction of cholera toxin and membrane GM1 ganglioside of small intestine. *Proc. Natl. Acad. Sci. U. S. A.* 72, 2520–2524.

Honerkamp-Smith, A. R., Cicuta, P., Collins, M. D., Veatch, S. L., den Nijs, M., Schick, M., Keller, S. L., 2008. Line tensions, correlation lengths, and critical exponents in lipid membranes near critical points. *Biophys. J.* 95, 236-246.

Horner, A., Akimov, S. A., Pohl, P., 2013. Long and short lipid molecules experience the same interleaflet drag in lipid bilayers. *Phys. Rev. Lett.* 110, 268101.

Iversen, L., Mathiasen, S., Larsen, J. B., Stamou, D., 2015. Membrane curvature bends the laws of physics and chemistry. *Nat. Chem. Biol.* 11, 822-825.

Jasenosky, L. D., Neumann, G., Lukashevich, I., Kawaoka, Y., 2001. Ebola virus VP40-induced particle formation and association with the lipid bilayer. *J. Virol.* 75, 5205-5214.

Jost, P. C., Griffith, O. H., Capaldi, R. A., Vanderkooi, G., 1973. Evidence for boundary lipid in membranes. *Proc. Natl. Acad. Sci. U. S. A.* 70, 480-484.

Justice, P. A., Sun, W., Li, Y., Ye, Z., Grigera, P. R., Wagner, R. R., 1995. Membrane vesiculation function and exocytosis of wild-type and mutant matrix proteins of vesicular stomatitis virus. *J. Virol.* 69, 3156-3160.

Kooijman, E. E., Tieleman, D. P., Testerink, C., Munnik, T., Rijkers, D. T., Burger, K. N., de Kruijff, B., 2007. An electrostatic/hydrogen bond switch as the basis for the specific interaction of phosphatidic acid with proteins. *J. Biol. Chem.* 282, 11356-11364.

Köse, M., 2017. GPCRs and EGFR - Cross-talk of membrane receptors in cancer. *Bioorg. Med. Chem. Lett.* 27, 3611-3620.

Kovács, T., Batta, G., Hajdu, T., Szabó, Á., Váradi, T., Zákány, F., Csomós, I., Szöllősi, J., Nagy, P., 2016. The dipole potential modifies the clustering and ligand binding affinity of ErbB proteins and their signaling efficiency. *Sci. Rep.* 6, 35850.

Kovács, T., Batta, G., Zákány, F., Szöllősi, J., Nagy, P., 2017. The dipole potential correlates with lipid raft markers in the plasma membrane of living cells. *J. Lipid Res.* 58, 1681-1691.

Kozlov, M. M., McMahon, H. T., Chernomordik, L. V., 2010. Protein-driven membrane stresses in fusion and fission. *Trends Biochem. Sci.* 35, 699-706.

Krabben, L., van Rossum, B. J., Jehle, S., Bocharov E., Lyukmanova, E. N., Schulga, A. A., Arseniev, A., Hucho, F., Oschkinat, H., 2009. Loop 3 of short neurotoxin II is an additional interaction site with membrane-bound nicotinic acetylcholine receptor as detected by solid-state NMR spectroscopy. *J. Mol. Biol.* 390, 662-671.

Ladisch, S., Kitada, S., Hays, E. F., 1987. Gangliosides shed by tumor cells enhance tumor formation in mice. *J. Clin. Invest.* 79, 1879-1882.

Landau, L. D., Lifshitz, E. M., 2013. *Statistical Physics*, Third Edition, Part 1: Volume 5 (Course of Theoretical Physics, Volume 5). Elsevier, ISBN 0080570461, 9780080570464.

Landau, M., Ben-Tal, N., 2008. Dynamic equilibrium between multiple active and inactive conformations explains regulation and oncogenic mutations in ErbB receptors. *Biochim. Biophys. Acta* 1785, 12–31.

Latham, T., Galarza, J. M., 2001. Formation of wild-type and chimeric influenza virus-like particles following simultaneous expression of only four structural proteins. *J. Virol.* 75, 6154-6165.

Lee, S. J., Li, Z., Litan, A., Yoo, S., Langhans, S. A., 2015. EGF-induced sodium influx regulates EGFR trafficking through HDAC6 and tubulin acetylation. *BMC Cell Biol.* 16, 24.

Lemmon, M. A., 2008. Membrane recognition by phospholipid-binding domains. *Nat. Rev. Mol. Cell. Biol.* 9, 99-111.

Lemmon, M. A., Schlessinger, J., 2010. Cell signaling by receptor tyrosine kinases. *Cell* 141, 1117–1134.

Leser, G. P., Lamb, R. A., 2017. Lateral organization of influenza virus proteins in the budozone region of the plasma membrane. *J. Virol.* 91, e02104-16.

Lesovoy, D. M., Bocharov, E. V., Lyukmanova, E. N., Kosinsky, Y. A., Shulepko, M. A., Dolgikh, D. A., Kirpichnikov, M. P., Efremov, R. G., Arseniev, A. S. 2009. Specific membrane binding of neurotoxin II can facilitate its delivery to acetylcholine receptor. *Biophys. J.* 97, 2089-2097.

Levine, T. P., Wiggins, C. A., Munro, S., 2000. Inositol phosphoryl-ceramide synthase is located in the Golgi apparatus of Saccharomyces cerevisiae. *Mol. Biol. Cell* 11, 2267–2281.

Li, E., Hristova, K., 2010. Receptor tyrosine kinase transmembrane domains: Function, dimer structure and dimerization energetics. *Cell Adh. Migr.* 4, 249-254.

Li, H., Papadopoulos, V., 1998. Peripheral-type benzodiazepine receptor function in cholesterol transport. Identification of a putative cholesterol recognition/interaction amino acid sequence and consensus pattern. *Endocrinology* 139, 4991-4997.

Li, L., Shi, X., Guo, X., Li, H., Xu, C., 2014. Ionic protein-lipid interaction at the plasma membrane: what can the charge do? *Trends Biochem. Sci.* 39, 130-140.

Liljeroos, L., Huiskonen, J. T., Ora, A., Susi, P., Butcher, S. J., 2011. Electron cryotomography of measles virus reveals how matrix protein coats the ribonucleocapsid within intact virions. *Proc. Natl. Acad. Sci. U. S. A.* 108, 18085-18090.

Lillemeier, B. F., Pfeiffer, J. R., Surviladze, Z., Wilson, B. S., Davis, M. M., 2006. Plasma membrane-associated proteins are clustered into islands attached to the cytoskeleton. *Proc. Natl. Acad. Sci. U. S. A.* 103, 18992-18997.

London, E., 2002. Insights into lipid raft structure and formation from experiments in model membranes. *Curr. Opin. Struct. Biol.* 12, 480-486.

Macdonald-Obermann, J. L., Pike, L. J., 2009. The intracellular juxtamembrane domain of the epidermal growth factor (EGF) receptor is responsible for the allosteric regulation of EGF binding. *J. Biol. Chem.* 284, 13570–13576.

MacKenzie, K. R., 2006. Folding and Stability of alpha-Helical Integral Membrane Proteins. *Chem. Rev.* 106 1931-1977.

Madsen, K. L., Bhatia, V. K., Gether, U., Stamou, D., 2010. BAR domains, amphipathic helices and membrane-anchored proteins use the same mechanism to sense membrane curvature. *FEBS Letters* 584, 1848-1855.

Martín, M., Albanesi, D., Alzari, P. M., De Mendoza, D., 2009. Functional in vitro assembly of the integral membrane bacterial thermosensor DesK. *Protein Expr. Purif.* 66, 39–45.

Martín-Nieto, J., Villalobo, A., 1998. The human epidermal growth factor receptor contains a juxtamembrane calmodulin-binding site. *Biochemistry* 37, 227–236.

Maruyama, I. N., 2015. Activation of transmembrane cell-surface receptors via a common mechanism? The "rotation model." *Bioessays* 37, 959-967.

Mattila, J. P., Shnyrova, A. V., Sundborger, A. C., Hortelano, E. R., Fuhrmans, M., Neumann, S., Müller M., Hinshaw J. E., Schmid S. L., Frolov, V. A., 2015. A hemi-fission intermediate links two mechanistically distinct stages of membrane fission. *Nature* 524, 109-113.

McIntosh, T. J., Vidal, A., Simon, S. A., 2003. Sorting of lipids and transmembrane peptides between detergent-soluble bilayers and detergent-resistant rafts. *Biophys. J.* 85, 1656-1666.

McMahon, H. T., Boucrot, E., 2015. Membrane curvature at a glance. *J. Cell. Sci.* 128, 1065-1070.

McMahon, H. T., Gallop, J. L., 2005. Membrane curvature and mechanisms of dynamic cell membrane remodelling. *Nature* 438, 590-596.

Mears, J. A., Fang, S., Leonard, M., Schmid, S. L., Milligan, R. A., J. E. Hinshaw Dyda, F., 2011. A pseudoatomic model of the dynamin polymer identifies a hydrolysis-dependent power stroke. *Cell* 147, 209-222.

Mineev, K. S., Panova, S. V., Bocharova, O. V., Bocharov, E. V., Arseniev, A. S., 2015. The membrane mimetic affects the spatial structure and mobility of EGFR transmembrane and juxtamembrane domains. *Biochemistry* 54, 6295-6298.

Molotkovskaya, I. M., Kholodenko, R. V., Molotkovsky, J. G., 2002. Influence of gangliosides on the IL-2–and IL-4–dependent cell proliferation. *Neurochem. Res.* 27, 761-770.

Monje-Galvan, V., Klauda, J. B., 2016. Peripheral membrane proteins: Tying the knot between experiment and computation. *Biochim. Biophys. Acta* 1858, 1584-1593.

Mouritsen, O. G., Bloom, M., 1984. Mattress model of lipid–protein interactions in membranes. *Biophys. J.* 46, 141–153.

Nadezhdin, K. D., Bocharova, O. V., Bocharov, E. V., Arseniev, A. S., 2011. Structural and dynamic study of the transmembrane domain of the amyloid precursor protein. *Acta Naturae* 3, 69-76.

Nadezhdin, K. D., Bocharova, O. V., Bocharov, E. V., Arseniev, A. S., 2012. Dimeric structure of transmembrane domain of amyloid precursor protein in micellar environment. *FEBS Lett.* 586, 1687-1692.

Nayak, D. P., Balogun, R. A., Yamada, H., Zhou, Z. H., Barman, S., 2009. Influenza virus morphogenesis and budding. *Virus Research* 143, 147-161.

Nojima, Y., Iwata, K., 2014. Viscosity heterogeneity inside lipid bilayers of single-component phosphatidylcholine liposomes observed with picosecond time-resolved fluorescence spectroscopy. *J. Phys. Chem. B* 118, 8631-8641.

Ou, S. I., Schrock, A. B., Bocharov, E. V., Klempner, S. J., Kawamura Haddad, C., Steinecker, G., Johnson, M., Gitlitz, B. J., Chung, J., Campregher, P. V., Ross, J. S., Stephens, P. J., Miller, V. A., Suh, J. H., Ali, S. M., Velcheti, V., 2017. HER2 transmembrane mutations (V659/G660) that stabilize homo- and hetero-dimerization are rare oncogenic drivers in lung adenocarcinoma that respond to afatinib. *J. Thorac. Oncol. 12*, 446-457.

Pantano, D. A., Moore, P. B., Klein, M. L., Discher, D. E., 2011. Raft registration across bilayers in a molecularly detailed model. *Soft Matter* 7, 8182-8191.

Perlmutter, J. D., Sachs, J. N., 2011. Interleaflet interaction and asymmetry in phase separated lipid bilayers: molecular dynamics simulations. *J. Am. Chem. Soc.* 133, 6563-6577.

Pike, L. J., 2006. Rafts defined: a report on the Keystone Symposium on Lipid Rafts and Cell Function. *J. Lip. Res.* 47, 1597-1598.

Pohl, C., Duprex, W. P., Krohne, G., Rima, B. K., Schneider-Schaulies, S., 2007. Measles virus M and F proteins associate with detergent-resistant membrane fractions and promote formation of virus-like particles. *J. Gen. Virol.* 88, 1243-1250.

Polozov, I. V., Bezrukov, L., Gawrisch, K., Zimmerberg, J., 2008. Progressive ordering with decreasing temperature of the phospholipids of influenza virus. *Nat. Chem. Biol.* 4, 248-255.

Porter, F. D., 2008. Smith–Lemli–Opitz syndrome: pathogenesis, diagnosis and management. *Eur. J. Hum. Genet.* 16, 535–541.

Pralle, A., Keller, P., Florin, E. L., Simons, K., Hörber, J. H., 2000. Sphingolipid-cholesterol rafts diffuse as small entities in the plasma membrane of mammalian cells. *J. Cell Biol.* 148, 997-1008.

Putzel, G. G., Schick, M., 2008. Phase behavior of a model bilayer membrane with coupled leaves. *Biophys. J.* 94, 869-877.

Putzel, G. G., Uline, M. J., Szleifer, I., Schick, M., 2011. Interleaflet coupling and domain registry in phase-separated lipid bilayers. *Biophys. J.* 100, 996-1004.

Rahman, M. S., Akhtar, N., Jamil, H. M., Banik, R. S., Asaduzzaman, S. M., 2015. TGF-β/BMP signaling and other molecular events: regulation of osteoblastogenesis and bone formation. *Bone Res.* 3, 15005.

Rinia, H. A., Snel, M. M., van der Eerden, J. P., de Kruijff, B., 2001. Visualizing detergent resistant domains in model membranes with atomic force microscopy. *FEBS Lett.* 501, 92-96.

Risselada, H. J., Marrink, S. J., 2008. The molecular face of lipid rafts in model membranes. *Proc. Natl. Acad. Sci. U. S. A.* 105, 17367-17372.

Ruigrok, R. W. H., Barge, A., Durrer, P., Brunner, J., Ma, K., Whittaker, G.R., 2000a. Membrane interaction of influenza virus M1 protein. *Virology* 267, 289-298.

Ruigrok, R. W. H., Schoehn, G., Dessen, A., Forest, E., Volchkov, V., Dolnik, O., Klenk, H. D., Weissenhorn, W., 2000b. Structural characterization and membrane binding properties of the matrix protein VP40 of Ebola virus. *J. Mol. Biol.* 300, 103-112.

Rukmini, R., Rawat, S. S., Biswas, S. C., Chattopadhyay, A., 2001. Cholesterol organization in membranes at low concentrations: effects of curvature stress and membrane thickness. *Biophys. J.* 81, 2122-2134.

Sagaram, U. S., El-Mounadi, K., Buchko, G. W., Berg, H. R., Kaur, J., Pandurangi, R. S., Smith, T. J., Shah, D. M., 2013. Structural and functional studies of a phosphatidic acid-binding antifungal plant defensin MtDef4: identification of an RGFRRR motif governing fungal cell entry. *PLoS One* 8, e82485.

Saita, E. A., Mendoza D., 2015. Thermosensing via transmembrane protein–lipid interactions. *Biochim. Biophys. Acta* 1848, 1757-1764.

Saito, T., Okada, S., Ohshima, K., Yamada, E., Sato, M., Uehara, Y., Shimizu, H., Pessin, J. E., Mori, M., 2004. Differential activation of epidermal growth factor (EGF) receptor downstream signaling pathways by betacellulin and EGF. *Endocrinology* 145, 4232–4243.

Saletti, D., Radzimanowski, J., Effantin, G., Midtvedt, D., Mangenot, S., Weissenhorn, W., Bassereau, P., Bally, M., 2017. The Matrix protein M1 from influenza C virus induces tubular membrane invaginations in an in vitro cell membrane model. *Sci. Rep.* 7, 40801.

Samsonov, A. V., Mihalyov, I., Cohen, F. S., 2001. Characterization of cholesterol-sphingomyelin domains and their dynamics in bilayer membranes. *Biophys. J.* 81, 1486-1500.

Sánchez-González, P., Jellali, K., Villalobo, A., 2010. Calmodulin-mediated regulation of the epidermal growth factor receptor. *FEBS J.* 277, 327-342.

Sarabipour, S., Hristova, K., 2016. Mechanism of FGF receptor dimerization and activation. *Nat. Commun.* 7, 10262.

Scheiffele, P., Rietveld, A., Wilk, T., Simons, K., 1999. Influenza viruses select ordered lipid domains during budding from the plasma membrane. *J. Biol. Chem.* 274, 2038-2044.

Schekman, R., Orci, L., 1996. Coat proteins and vesicle budding. *Science* 271, 1526-1533.

Schmid, S. L., Frolov, V. A., 2011. Dynamin: functional design of a membrane fission catalyst. *Annu. Rev. Cell Dev. Biol.* 27, 79-105.

Sengupta, P., Bosis, E., Nachliel, E., Gutman, M., Smith, S. O., Mihályné, G., Zaitseva, I., McLaughlin, S., 2009. EGFR juxtamembrane domain, membranes, and calmodulin: kinetics of their interaction. *Biophys. J.* 96, 4887–4895.

Sha, B., Luo, M., 1997. Structure of a bifunctional membrane-RNA binding protein, influenza virus matrix protein M1. *Nat. Struct. Mol. Biol.* 4, 239-244.

Sharonov, G. V., Balatskaya, M. N., Tkachuk, V. A., 2016. Glycosylphosphatidylinositol-anchored proteins as regulators of cortical cytoskeleton. *Biochemistry (Mosc)* 81, 636–650.

Sharonov, G. V., Bocharov, E. V., Kolosov, P. M., Astapova, M. V., Arseniev, A. S., Feofanov, A. V., 2014. Point mutations in dimerization motifs of transmembrane domain stabilize active or inactive state of the EphA2 receptor tyrosine kinase. *J. Biol. Chem.* 289, 14955-14964.

Sharpe, H. J., Stevens, T. J., Munro, S., 2010. A comprehensive comparison of transmembrane domains reveals organelle-specific properties. *Cell* 142, 158-169.

Shi, Z., Baumgart, T., 2015. Membrane tension and peripheral protein density mediate membrane shape transitions. *Nat. Comm.* 6, 5974.

Shilova, L. A., Knyazev, D. G., Fedorova, N. V., Shtykova, E. V., Batishchev, O. V., 2017. Study of adsorption of Influenza virus matrix protein M1 on lipid membranes by the technique of fluorescent probes. *Biochemistry (Moscow), Supplement Series A: Membrane and Cell Biology* 11, 225-230.

Shnyrova, A. V., Bashkirov, P. V., Akimov, S. A., Pucadyil, T. J., Zimmerberg, J., Schmid, S. L., Frolov, V. A., 2013. Dynamin-driven membrane fission reveals catalytic principle of membrane remodeling. *Science* 339, 1433-1436.

Shnyrova, A. V., Ayllon, J., Mikhalyov, I. I., Villar, E., Zimmerberg, J., Frolov, V. A., 2007. Vesicle formation by self-assembly of membrane-bound matrix proteins into a fluidlike budding domain. *J. Cell. Biol.* 179, 627-633.

Shtykova, E. V., Baratova, L. A., Fedorova, N. V., Radyukhin, V. A., Ksenofontov, A. L., Volkov, V. V., Shishkov, A. V., Dolgov, A. A., Shilova, L. A., Batishchev O. V., Jeffries, C. M., Svergun D. I., 2013. Structural analysis of influenza A virus matrix protein M1 and its self-assemblies at low pH. *PloS one* 8, e82431.

Sidransky, E., 2012. Gaucher disease: insights from a rare Mendelian disorder. *Discov. Med.* 14, 273-281.

Simons, K., Ikonen, E., 1997. Functional rafts in cell membranes. *Nature* 387, 569-572.

Simunovic, M., Evergren, E., Golushko, I., Prévost, C., Renard, H. F., Johannes, L., McMahon, H. T., Lorman, V., Voth, G. A., Bassereau, P., 2016. How curvature-generating proteins build scaffolds on membrane nanotubes. *Proc. Natl. Acad. Sci. U. S. A.* 113, 11226-11231.

Slater, J. L., Huang, C. H., 1988. Interdigitated bilayer membranes. *Prog. Lip. Res.* 27, 325-359.

Snead, W. T., Hayden, C. C., Gadok, A. K., Zhao, C., Lafer, E. M., Rangamani, P., Stachowiak, J. C., 2017. Membrane fission by protein crowding. *Proc. Natl. Acad. Sci. U. S. A.* 114, E3258-E3267.

Stachowiak, J. C., Hayden, C. C., Sasaki, D. Y., 2010. Steric confinement of proteins on lipid membranes can drive curvature and tubulation. *Proc. Natl. Acad. Sci. U. S. A.* 107, 7781-7786.

Staneva, G., Osipenko, D. S., Galimzyanov, T. R., Pavlov, K. V., Akimov, S. A., 2016. Metabolic precursor of cholesterol causes formation of chained aggregates of liquid-ordered domains. *Langmuir* 32, 1591-1600.

Suzuki, K. G., 2015. New insights into the organization of plasma membrane and its role in signal transduction. *Int. Rev. Cell. Mol. Biol.* 317, 67-96.

Usery, R. D., Enoki, T. A., Wickramasinghe, S. P., Weiner, M. D., Tsai, W. C., Kim, M. B., Wang, S., Torng, T. L., Ackerman, D. G., Heberle,

F. A., Katsaras, J., Feigenson, G. W., 2017. Line tension controls liquid-disordered+ liquid-ordered domain size transition in lipid bilayers. *Biophys. J.* 112, 1431-1443.

Vanni, S., Hirose, H., Barelli, H., Antonny, B., Gautier, R., 2014. A sub-nanometre view of how membrane curvature and composition modulate lipid packing and protein recruitment. *Nat. Commun.* 5, 4916.

Veatch, S. L., Cicuta, P., Sengupta, P., Honerkamp-Smith, A., Holowka, D., Baird, B., 2008. Critical fluctuations in plasma membrane vesicles. *ACS Chem. Biol.* 3, 287-293.

Veatch, S. L., Polozov, I. V., Gawrisch, K., Keller, S. L., 2004. Liquid domains in vesicles investigated by NMR and fluorescence microscopy. *Biophys. J.* 86, 2910-2922.

Veit, M., Thaa, B., 2011. Association of influenza virus proteins with membrane rafts. *Adv. Virol.* 370606.

Volynsky, P. E., Polyansky, A. A., Bocharov, E. V., Efremov, R. G. 2013. Role of dimerization efficiency of transmembrane domains in activation of fibroblast growth factor receptor 3. *J. Am. Chem. Soc.* 135, 8105–8108.

Wang, Y., Gao, J., Guo, X., Tong, T., Shi, X., Li, L., Qi, M., Wang, Y., Cai, M., Jiang, J., Xu, C., Ji, H., Wang, H., 2014. Regulation of EGFR nanocluster formation by ionic protein-lipid interaction. *Cell Res.* 24, 959-976.

Waters, M. J., Brooks, A. J., 2015. JAK2 activation by growth hormone and other cytokines. *Biochem. J.* 466 1-11.

White, J. M., Whittaker, G. R., 2016. Fusion of enveloped viruses in endosomes. *Traffic* 17, 593-614.

Wijesinghe, K. J., Urata, S., Bhattarai, N., Kooijman, E. E., Gerstman, B. S., Chapagain, P. P., Li, S., Stahelin, R. V., 2017. Detection of lipid-induced structural changes of the Marburg virus matrix protein VP40 using hydrogen/deuterium exchange-mass spectrometry. *J. Biol. Chem.* 292, 6108-6122.

Williamson, J. J., Olmsted, P. D., 2015. Registered and antiregistered phase separation of mixed amphiphilic bilayers. *Biophys. J.* 108, 1963-1976.

Wilson, K. J., Gilmore, J. L., Foley, J., Lemmon, M. A., Riese, D. J., 2009. Functional selectivity of EGF family peptide growth factors: Implications for cancer. *Pharmacol. Ther.* 122, 1–8.

Ye, Z. P., Pal, R., Fox, J. W., Wagner, R. R., 1987. Functional and antigenic domains of the matrix (M1) protein of influenza A virus. *J. Virol.* 61, 239-246.

Zhang, W., Zheng, W., Toh, Y., Betancourt-Solis, M. A., Tu, J., Fan, Y., Vakharia, V. N., Liu, J., McNew, J. A., Jin, M., Tao, Y. J., 2017. Crystal structure of an orthomyxovirus matrix protein reveals mechanisms for self-polymerization and membrane association. *Proc. Natl. Acad. Sci. U. S. A.* 114, 8550-8555.

Zimmerberg, J., Kozlov, M. M., 2006. How proteins produce cellular membrane curvature. *Nat. Rev. Mol. Cell Biol.* 7, 9-19.

In: Protein-Lipid Interactions
Editor: Angel Catalá ISBN: 978-1-53613-125-3

Chapter 7

IMPORTANT ASPECTS IN THE KNOWLEDGE OF PROTEIN–LIPID INTERACTIONS FROM 1900 UP TO 2017

Angel Catalá[*]
Instituto de Investigaciones Fisicoquímicas Teóricas y Aplicadas, (INIFTA-CCT La Plata-CONICET), Facultad de Ciencias Exactas, Universidad Nacional de La Plata, Argentina

ABSTRACT

In the last five decades, the significance of protein–lipid interactions in biological systems began to be widely explored; the most important aspects in the knowledge of protein–lipid interactions from 1969 up to now are described in this chapter.

INTRODUCTION

Biological systems contain many different membrane compartments with characteristic shapes, lipid compositions, and dynamics. A huge portion of cytoplasmic proteins connect with these membrane

[*] Corresponding Author Email: catala@inifta.unlp.edu.ar.

compartments. Such protein-lipid interactions, which control the subcellular localizations and activities of peripheral membrane proteins, are essentially important for a diversity of cell biological processes ranging from cytoskeletal dynamics and membrane trafficking to intracellular signaling. Equally, many membrane-associated proteins can modulate the shape, lipid composition, and dynamics of cellular membranes. Determining the exact mechanisms by which these proteins interact with membranes will be indispensable to consider their biological functions. With this objective in mind, I provide a brief introduction to selected biochemical articles that can be checked with to study protein-lipid interactions.

The Significance of Protein–Lipid Interactions in Biological Systems

In the last five decades, the significance of protein–lipid interactions in biological systems began to be widely explored; the most important aspects in the knowledge of protein–lipid interactions from 1969 up to now are listed below:

- **Physical studies of lipid-lipid and lipid-protein interactions.** Chapman D. *Lipids*. 1969 4: 251-260. Review
- **Physical properties of staphylococcal alpha-toxin and aspects of alpha-toxin membrane interactions.** Arbuthnott JP, Freer JH, McNiven AC. *Contrib Microbiol Immunol*. 1973; 1: 285-97. Review.
- **Lipid-protein interactions in the structure of biological membranes.** Lenaz G. *Subcell Biochem*. 1974 3: 167-248. Review
- **Fatty acid binding to plasma albumin.** Spector AA. J *Lipid Res*. 1975 3:165-167 In this review Spector described the available information about fatty acid binding to plasma albumin.
- **Modes of lipid-protein interactions in biomembranes.** Wallach DF, Bieri V, Verma SP, Schmidt-Ullrich R. *Ann N Y Acad Sci*. 1975; 264: 142-160. Review

- **Protein-liposome interactions and their relevance to the structure and function of cell membranes.** Kimelberg HK. *Mol Cell Biochem.* 1976; 3: 171-190. Review
- **Protein-lipid interactions.** Gennis RB. *Annu Rev Biophys Bioeng.* 1977; 6:195-238. Review.
- **Lipid-protein interactions in the plasma lipoproteins.** Morrisett JD, Jackson RL, Gotto AM Jr. *Biochim Biophys Acta.* 1977; 472: 93-133. Review. The purpose of this review has been to discuss new information about the mechanism of lipid and apoprotein interaction in the plasma lipoproteins.
- **Protein-lipid interactions and the role of water.** Edelhoch H. Horiz *Biochem Biophys.* 1978; 5:241-280. Review. This study demonstrates that the rigidity of the three-dimensional structure of a native protein is dependent on the network of hydrogen-bonded groups which provide the scaffolding for the other interactions. The structure is stabilized by the hydrophobic interactions of the nonpolar side chains. The latter are formed by the very unfavorable entropy change that occurs in water but not in less-polar solvents.
- **Intrinsic protein--lipid interactions. Physical and biochemical evidence.** Chapman D, Gómez-Fernández JC, Goñi FM. *FEBS Lett.* 1979; 98: 211-223. Review The fact that an intrinsic protein may perturb its surrounding lipid is clear. The type of perturbation depends, however, on the concentration of protein within the lipid bilayer. At low protein concentrations, an ordering of lipid chains can occur whilst at high protein concentrations a disordering effect occurs. At these high concentrations the perturbation of the lipid polar group packing has to be considered. The time scale appropriate to the particular physical technique used to study the protein-lipid perturbation must be related to the time scale of the phenomenon being examined. Concepts such as boundary-layer lipids or annulus lipids, implying long-lived shells of lipid controlling enzyme activity, and only slowly exchanging with fluid lipid, are not supported by recent NMR experiments. The ESR data showing an 'immobile component' should be reconsidered in

terms of other effects, e.g., due to high microviscosity of trapped lipid, the specific attachment of the ESR probe to the protein, or lipid inside eutectic mixtures. The formation of aggregates or eutectic mixtures of high protein to lipid content is an important consideration when reconstituted systems and membranes are studied at temperatures below the Tc transition temperature of the pure lipid. Effect of phase transitions on the interaction of peptides and proteins with phospholipids.

- Epand RM, Surewicz WK *Can J Biochem Cell Biol.* 1984 62:1167-1173. The effect of proteins on lipid order and motion above and below the phase transition is discussed and the current state of knowledge on this topic is briefly reviewed. Nuclear magnetic resonance methods to characterize lipid-protein interactions at membrane surfaces.
- Watts A. J *Bioenerg Biomembr.* 1987; 19:625-653. Review. A comprehensive review of the literature will not be presented here. However, protein-lipid molecular specificity in membranes, peptide-induced lateral separation, and the ionization behavior of deuterated phospholipids and peripheral proteins will all be demonstrated predominantly using deuterium NMR methods. Some suggestions for future work are also presented.
- **Interactions of lipases with lipid monolayers. Facts and questions.** Piéroni G, Gargouri Y, Sarda L, Verger R *Adv Colloid Interface Sci.* 1990; 32: 341-378. Among the proteins, lipolytic enzymes provide a valuable model for studying protein-lipid interactions. Lipases having a catalytic action which is strictly dependent upon the presence of a lipid interface were used in the present study in order to gain better insight into protein-lipid interactions. Most of the data presented here were obtained using the monolayer technique, by recording (either independently or simultaneously) the lipolytic activity, the amount of protein adsorbed to the lipid monolayer, and the surface pressure variations following protein adsorption. Several non-enzymatic

proteins were used as controls in order to determine how lipase behavior differs from that of other proteins.

- **The effects of non-lamellar forming lipids on membrane protein-lipid interactions.** Stubbs CD, Slater SJ. *Chem Phys Lipids*. 1996; 81:185-195. Review. The role of lipid polymorphism in the regulation of membrane-associated protein function is examined, based on recent studies which showed that changes in the levels of phosphatidylethanolamine (PE), cholesterol and phospholipid unsaturation, modulate the activity of the key signal transduction enzyme, protein kinase C (PKC). It is shown that effects of membrane compositional changes on PKC activity involve a perturbation of protein-lipid interactions with the head group region rather than with the hydrophobic interior of the bilayer. A key determinant in the perturbation of these interactions is suggested to be an elastic curvature energy, termed curvature stress, which results from the unfavorable packing of non-lamellar forming lipids in a planar bilayer.
- **Theoretical analysis of protein organization in lipid membranes.** Gil T, Ipsen JH, Mouritsen OG, Sabra MC, Sperotto MM, Zuckermann MJ. *Biochim Biophys Acta*. 1998; 1376: 245-266. Review. The fundamental physical principles of the lateral organization of trans-membrane proteins and peptides as well as peripheral membrane proteins and enzymes are considered from the point of view of the lipid-bilayer membrane, its structure, dynamics, and cooperative phenomena. Based on a variety of theoretical considerations and model calculations, the nature of lipid-protein interactions is considered both for a single protein and an assembly of proteins that can lead to aggregation and protein crystallization in the plane of the membrane.
- **Lipid polymorphism and protein-lipid interactions.** Epand RM. *Biochim Biophys Acta*. 1998;1376:353-368. Non-lamellar-forming lipids play an important role in determining the physical properties of membranes. They affect the activity of membrane proteins and peptides. In addition, peptides which lyse membranes as well as

those which promote membrane fusion facilitate the formation of non-lamellar phases, either micelles, cubic or hexagonal phases. The relationship of these diverse effects on membrane curvature is discussed in relation to the function of certain peptides and proteins.

- **Lipid monolayers: why use half a membrane to characterize protein-membrane interactions?** Brockman H. *Curr Opin Struct Biol.* 1999; 9: 438-443. Variants of membrane-active proteins and peptides are increasingly available through synthesis and molecular engineering. When determining the effects of structural changes upon the interaction of these proteins with lipid membranes, monomolecular films of lipids at the air-water interface have significant advantages over bilayers and other lipid dispersions. In the past year, a variety of protein-lipid interactions has been characterized successfully using relatively simple surface measurements.
- **Is the protein/lipid hydrophobic matching principle relevant to membrane organization and functions?** Dumas F, Lebrun MC, Tocanne JF. *FEBS Lett.* 1999; 458: 271-277. Biological membranes are complex and well-organized multimolecular assemblies composed of a wide variety of protein and lipid molecular species. If such a diversity in protein and lipid polar head group structures may easily be related to a large panel of functions, the wide dispersion in acyl chain length and structure which the lipids display is more difficult to understand. It is not required for maintaining bilayer assembly and fluidity. Direct information on the lateral distribution of these various molecular species, on their potential specificity for interaction between themselves and with proteins and on the functional implications of these interactions is also still lacking. Because hydrophobic interactions play a major role in stabilizing membrane structures, we suggest considering the problem from the point of view of the matching of the hydrophobic surface of proteins by the acyl chains of the lipids.

- **Protein myristoylation in protein-lipid and protein-protein interactions.** Taniguchi H. *Biophys Chem.* 1999; 82: 129-137. Various proteins in signal transduction pathways are myristoylated. Although this modification is often essential for the proper functioning of the modified protein, the mechanism by which the modification exerts its effects is still largely unknown. Here we discuss the roles played by protein myristoylation, in both protein-lipid and protein-protein interactions. Myristoylation is involved in the membrane interactions of various proteins, such as MARCKS and endothelial NO synthase. The intermediate hydrophobic nature of the modification plays an important role in the reversible membrane anchoring of these proteins. The anchoring is strengthened by a basic amphiphilic domain that works as a switch for the reversible binding. Protein myristoylation is also involved in protein-protein interactions, which are regulated by the interplay between protein phosphorylation, calmodulin binding, and membrane phospholipids.
- **Structure determination of membrane-associated proteins from nuclear magnetic resonance data.** Baleja JD. *Anal Biochem.* 2001; 288: 1-15. This Review covers the delineation and optimization of protein-lipid systems for study using solution-state NMR spectroscopy. The first half presents the necessary background for a membrane protein biochemist to initiate collaboration with an NMR spectroscopist. The second half provides guidelines for the spectroscopist on data collection, analysis for obtaining conformational information, and structure generation and assessment. Although the emphasis is on the study of peptides in detergent micelles, methods are outlined for larger membrane-associated proteins and for use of other solubilizing agents.
- **Probing the interface between membrane proteins and membrane lipids by X-ray crystallography.** Fyfe PK, McAuley KE, Roszak AW, Isaacs NW, Cogdell RJ, Jones MR. *Trends Biochem Sci.* 2001; 26: 106-112. Biological membranes are

composed of a complex mixture of lipids and proteins, and the membrane lipids support several key biophysical functions, in addition to their obvious structural role. Recent results from X-ray crystallography are shedding new light on the precise molecular details of the protein-lipid interface.

- **Protein-lipid interactions in food systems: A review.** Alzagtat AA, Alli I. *Int J Food Sci Nutr*. 2002; 53: 249-260. Proteins and lipids, both individually or as complexes, play important functional roles in foods. Since the 1970s food scientists have devoted attention to the nature of these interactions and particularly to their effects on functional characteristics of protein-based foods. Previously, most of the published work was devoted to the biochemical aspects of protein-lipid interactions in biological systems. This article reviews the protein-lipid interactions of both naturally occurring protein-lipid complexes and protein-lipid complexes formed by induced interactions in foods and food products. The physicochemical characteristics of known protein-lipid complexes, the nature of binding which results in formation of these complexes and the effect of the interactions on food functionality are reviewed.
- **Interaction of membrane-spanning proteins with peripheral and lipid-anchored membrane proteins: perspectives from protein-lipid interactions (Review).** Marsh D, Horváth LI, Swamy MJ, Mantripragada S, Kleinschmidt JH. *Mol Membr Biol*. 2002; 19: 247-255. Studies of lipid-protein interactions in double-reconstituted systems involving both integral and peripheral or lipid-anchored proteins are reviewed. Membranes of dimyristoyl phosphatidylglycerol containing either myelin proteolipid protein or cytochrome c oxidase were studied. The partner peripheral proteins bound to these membranes were myelin basic protein or cytochrome c, respectively. In addition, the interactions between the myelin proteolipid protein and avidin that was membrane-anchored by binding to N-biotinyl phosphatidylethanolamine were studied in dimyristoyl phosphatidylcholine membranes.

- **Protein-lipid interplay in fusion and fission of biological membranes.** Chernomordik LV, Kozlov MM. *Annu Rev Biochem.* 2003; 72:175-207. Disparate biological processes involve fusion of two membranes into one and fission of one membrane into two. To formulate the possible job description for the proteins that mediate remodeling of biological membranes, we analyze the energy price of disruption and bending of membrane lipid bilayers at the different stages of bilayer fusion. We briefly consider some proteins involved in fusion and fission, and the dependence of remodeling on the lipid composition of the membranes. The specific hypothetical mechanisms by which the proteins can lower the energy price of the bilayer rearrangement are discussed in light of the experimental data and the requirements imposed by the elastic properties of the bilayer.
- **The protein-lipid interface: perspectives from magnetic resonance and crystal structures.** Marsh D, Páli T. *Biochim Biophys Acta.* 2004; 1666:118-141. Lipid-protein interactions in membranes are dynamic, and consequently are well studied by magnetic resonance spectroscopy. More recently, lipids associated with integral membrane proteins have been resolved in crystals by X-ray diffraction, mostly at cryogenic temperatures. The conformation and chain ordering of lipids in crystals of integral proteins are reviewed here and are compared and contrasted with results from magnetic resonance and with the crystal structures of phospholipid bilayers. Various aspects of spin-label magnetic resonance studies on lipid interactions with single integral proteins are also reviewed: specificity for phosphatidylcholine, competition with local anaesthetics, oligomer formation of single transmembrane helices, and protein-linked lipid chains. Finally, the interactions between integral proteins and peripheral or lipid-linked proteins, as reflected by the lipid-protein interactions in double reconstitutions, are considered.
- **Structural basis for lipid modulation of nicotinic acetylcholine receptor function.** Barrantes FJ. *Brain Res Brain Res Rev.* 2004;

47: 71-95. The nicotinic acetylcholine receptor (AChR) is the archetype molecule in the superfamily of ligand-gated ion channels (LGIC). Members of this superfamily mediate fast intercellular communication in response to endogenous neurotransmitters. This review is focused on the structural and functional crosstalk between the AChR and lipids in the membrane microenvironment, and the modulation exerted by the latter on ligand binding and ion translocation.

- **Specific protein-lipid interactions in membrane proteins.** Hunte C. *Biochem Soc Trans*. 2005; 33: 938-942. Many membrane proteins selectively bind defined lipid species. This specificity has an impact on correct insertion, folding, structural integrity and full functionality of the protein. How are these different tasks achieved? Recent advances in structural research of membrane proteins provide new information about specific protein-lipid interactions. Tightly bound lipids in membrane protein structures are described and general principles of the binding interactions are deduced. The importance of lipid binding for vertical positioning and tight integration of proteins in the membrane, for assembly and stabilization of oligomeric and multisubunit complexes, for supercomplexes, as well as for functional roles are pointed out.
- **Domains in biological membranes.** Lindner R, Naim HY. *Exp Cell Res*. 2009; 315: 2871-2878/ Our current view on biological membranes has gradually evolved from the influential fluid mosaic model of the early 1970s to a distinctively more complex picture. Biological membranes are now assumed to encompass multiple membrane domains and a plethora of protein-lipid and protein-protein interactions that compartmentalize and temporarily order what has originally been envisioned to be mostly random. In this minireview, we will first highlight some structural principles that govern membrane domain formation and permit a classification of membrane domains. We will then focus on the still controversial issue of lipid-based membrane domains, or lipid rafts, and discuss recent advances in detecting these enigmatic structures in living

cells. Finally we will evaluate biochemical approaches to characterize lipid rafts and discuss their contribution to the emerging topic of lipid raft diversity.

- **Protein lipid modifications in signaling and subcellular targeting.** Sorek N, Bloch D, Yalovsky S. *Curr Opin Plant Biol.* 2009; 12: 714-720. Classically perceived as means for recruiting proteins to the membranes, protein lipid modifications are known today to play diverse roles in subcellular targeting, protein-protein interactions and signaling. This review focuses on three protein lipid modifications: prenylation, S-acylation and N-myristoylation and attempts to provide an up-to-date view of their function by focusing on several model proteins.
- **Determinants of specificity at the protein-lipid interface in membranes.** Ernst AM, Contreras FX, Brügger B, Wieland F. *FEBS Lett.* 2010; 584:1713-1720. The complexity of pro- and eukaryotic lipidomes is increasingly appreciated mainly owing to the advance of mass spectrometric methods. Biophysical approaches have revealed that the large number of lipid classes and molecular species detected have implications for the self-organizing potential of biological membranes, resulting in the formation of lateral heterogeneous phases. How membrane proteins are able to adapt specifically to their surrounding heterogeneous matrix, and whether this environment affects protein targeting and function, is therefore a matter of particular interest. Here, we review specific protein-lipid interactions, focusing on the molecular mechanisms that determine specificity at the protein-lipid interface, and on membrane proteins that require lipids as cofactors for their architecture and function.
- **Quantification of protein-lipid selectivity using FRET.** Loura LM, Prieto M, Fernandes F. *Eur Biophys J.* 2010; 39: 565-578. Membrane proteins exhibit different affinities for different lipid species, and protein-lipid selectivity regulates the membrane composition in close proximity to the protein, playing an important role in the formation of nanoscale membrane heterogeneities. The

sensitivity of Förster resonance energy transfer (FRET) for distances of 10 A up to 100 A is particularly useful to retrieve information on the relative distribution of proteins and lipids in the range over which protein-lipid selectivity is expected to influence membrane composition. Several FRET-based methods applied to the quantification of protein-lipid selectivity are described herein, and different formalisms applied to the analysis of FRET data for particular geometries of donor-acceptor distribution are critically assessed.

- **A lipid matrix model of membrane raft structure.** Quinn PJ. *Prog Lipid Res.* 2010; 49: 390-406. The lipid matrix model provides a plausible molecular mechanism to explain how membrane rafts are formed and perform their various functions. Specificity in creation of particular raft platforms resides in the N-acyl fatty acid substituents of the sphingolipids and the sugar residues of glycosphingolipids. Organization into signal transducing elements is bought about by the molecular lattice formed by the precise arrangement of glycosphingolipid and phospholipids in the quasi-crystalline arrays. Thus the stoichiometry of the lattice itself creates the framework for efficient coupling between the two halves of the bilayer.
- **A simple guide to biochemical approaches for analyzing protein-lipid interactions.** Zhao H, Lappalainen P. *Mol Biol Cell.* 2012; 23: 2823-2830. In this Technical Perspective, we provide a brief introduction to selected biochemical methods that can be applied to study protein-lipid interactions. We also discuss how important it is to choose proper lipid composition, type of model membrane, and biochemical assay to obtain reliable and informative data from the lipid-interaction mechanism of a protein of interest.
- **Nanodiscs as a new tool to examine lipid-protein interactions.** Schuler MA, Denisov IG, Sligar SG. *Methods Mol Biol.* 2013; 974: 415-433. Nanodiscs are self-assembled discoidal fragments of lipid bilayers 8-16 nm in diameter, stabilized in solution by two

amphipathic helical scaffold proteins. As stable and highly soluble membrane mimetics with controlled lipid composition and ability to add affinity tags to the scaffold protein, nanodiscs represent an attractive model system for solubilization, isolation, purification, and biophysical and biochemical studies of membrane proteins. Special attention is paid to the opportunities afforded by nanodisc system for the detailed studies of the role of different lipid properties and protein-lipid interactions in the functional behavior of membrane proteins.

- **Solid-state NMR spectroscopy to study protein-lipid interactions.** Huster D. *Biochim Biophys Acta*. 2014; 1841:1146-1160. Methods that detect the influence of proteins on the membrane as well as direct lipid-protein interactions have been developed and are reviewed here. Examples for solid-state NMR studies on the interaction of Ras proteins, the antimicrobial peptide protegrin-1, the G protein-coupled receptor rhodopsin, and the K(+) channel KcsA are discussed. This article is part of a Special Issue entitled Tools to study lipid functions.
- **Dynamic regulation of lipid-protein interactions.** Martfeld AN, Rajagopalan V, Greathouse DV, Koeppe RE. *Biochim Biophys Acta*. 2015; 1848:1849-1859. We review the importance of helix motions for the function of several important categories of membrane proteins and for the properties of several model molecular systems. Peptides which initially associate with lipid-bilayer membrane surfaces may undergo dynamic transitions from surface-bound to tilted-transmembrane orientations, sometimes accompanied by changes in the molecularity, formation of a pore or, more generally, the activation of biological function. For single-span membrane proteins, such as the tyrosine kinases, interplay between juxtamembrane and transmembrane domains are likely to be crucial for the regulation of dimer assembly that in turn is associated with the functional responses to external signals. Additionally, we note that experiments with designed single-span transmembrane helices offer fundamental insights into the

molecular features that govern protein-lipid interactions. This article is part of a Special Issue entitled: Lipid-protein interactions.

- **Electrostatically driven lipid-protein interaction: Answers from FRET.** Fernandes F, Coutinho A, Prieto M, Loura LM. *Biochim Biophys Acta.* 2015; 1848: 1837-1848. Electrostatics governs the association of a large number of proteins with cellular membranes. In some cases, these proteins present specialized lipid-binding modules or membrane targeting domains while in other cases association is achieved through nonspecific interaction of unstructured clusters of basic residues with negatively charged lipids. In this review we present and discuss the basic formalisms of both hetero- and homo-FRET pertinent to the most commonly encountered problems in lipid-protein interaction studies and highlight some examples of implementations of different FRET methodologies to characterize lipid/protein systems in which electrostatic interactions play a crucial role. This article is part of a Special Issue entitled: Lipid-protein interactions.
- **Lipid-protein interplay and lateral organization in biomembranes.** Nyholm TK. *Chem Phys Lipids.* 2015; 189: 48-55. The distribution of lipids and membrane proteins within and between cellular membranes is carefully regulated. This sorting of lipids and proteins results in compositional differences between membrane compartments, bilayer leaflets, and lateral domains in the bilayer plane. The lateral organization of lipids and proteins has proven challenging to investigate, and the driving mechanisms remains unclear. Lipid self-organization has an essential role in the formation of lateral membrane structure, but the role of transmembrane proteins is not well known. This review focuses on how lateral lipid structure can affect the lateral distribution of proteins and how proteins could influence the organization of lipids in biomembranes.
- **The interactions of peripheral membrane proteins with biological membranes.** Whited AM, Johs A. *Chem Phys Lipids.* 2015; 192: 51-59. The interactions of peripheral proteins with

membrane surfaces are critical to many biological processes, including signaling, recognition, membrane trafficking, cell division and cell structure. On a molecular level, peripheral membrane proteins can modulate lipid composition, membrane dynamics and protein-protein interactions. Biochemical and biophysical studies have shown that these interactions are in fact highly complex, dominated by several different types of interactions, and have an interdependent effect on both the protein and membrane. Here the authors examine three major mechanisms underlying the interactions between peripheral membrane proteins and membranes: electrostatic interactions, hydrophobic interactions, and fatty acid modification of proteins.

- **The systematic analysis of protein-lipid interactions comes of age.** Saliba AE, Vonkova I, Gavin AC. *Nat Rev Mol Cell Biol.* 2015; 16: 753-761. Lipids tailor membrane identities and function as molecular hubs in all cellular processes. However, the ways in which lipids modulate protein function and structure are poorly understood and still require systematic investigation. In this article, the authors summarize pioneering technologies, including lipid-overlay assays, lipid pull-down assays, affinity-purification lipidomics and the liposome microarray-based assay (LiMA) that will enable protein-lipid interactions to be deciphered on a systems level. The authors discuss how these technologies can be applied to the charting of system-wide networks and to the development of new pharmaceutical.
- **Protein lipid modifications--More than just a greasy ballast.** Hentschel A, Zahedi RP, Ahrends R. *Proteomics.* 2016; 16:759-782. Covalent lipid modifications of proteins are crucial for regulation of cellular plasticity, since they affect the chemical and physical properties and therefore protein activity, localization, and stability. Most recently, lipid modifications on proteins are increasingly attracting important regulatory entities in diverse signaling events and diseases. In all cases, the lipid moiety of modified proteins is essential to allow water-soluble proteins to

strongly interact with membranes or to induce structural changes in proteins that are critical for elemental processes such as respiration, transport, signal transduction, and motility. Therefore, this review aims to give a comprehensive and detailed overview about enzymatic and nonenzymatic lipidation events, the authors report here their role in cellular biology, discuss their relevancy for diseases, and describe so far available bioanalytical strategies to analyze this highly challenging type of modification.

- **Atomic-level description of protein-lipid interactions using an accelerated membrane model.** Baylon JL, Vermaas JV, Muller MP, Arcario MJ, Pogorelov TV, Tajkhorshid E. *Biochim Biophys Acta.* 2016; 1858:1573-1783. Peripheral membrane proteins are structurally diverse proteins that are involved in fundamental cellular processes. The activity of these proteins is frequently modulated through their interaction with cellular membranes, and as a result techniques to study the interfacial interaction between peripheral proteins and the membrane are in high demand. To increase lipid dynamics while preserving the atomistic detail of protein-lipid interactions, in the highly mobile membrane-mimetic (HMMM) model the membrane core is replaced by an organic solvent, while short-tailed lipids provide a nearly complete representation of natural lipids at the organic solvent/water interface. Here, the authors present a brief introduction and a summary of recent applications of the HMMM to study different membrane proteins, complementing the experimental characterization of the presented systems, and we offer a perspective of future applications of the HMMM to study other classes of membrane proteins.
- **Fatty acylation of proteins: The long and the short of it.** Resh MD. *Prog Lipid Res.* 2016; 63: 120-131. Long, short and medium chain fatty acids are covalently attached to hundreds of proteins. Each fatty acid confers distinct biochemical properties, enabling fatty acylation to regulate intracellular trafficking, subcellular localization, protein-protein and protein-lipid interactions.

Myristate and palmitate represent the most common fatty acid modifying groups. Enzymes that remove N-terminal myristoyl-glycine or myristate from lysines have now been identified. DHHC proteins catalyze S-palmitoylation, but the mechanisms that regulate substrate recognition by individual DHHC family members remain to be determined. New studies continue to reveal thioesterases that remove palmitate from S-acylated proteins. Another area of rapid expansion is fatty acylation of the secreted proteins hedgehog, Wnt and Ghrelin, by Hhat, Porcupine and GOAT, respectively. Understanding how these membrane bound O-acyl transferases recognize their protein and fatty acyl CoA substrates is an active area of investigation, and is punctuated by the finding that these enzymes are potential drug targets in human diseases.

- **The two sides of a lipid-protein story.** Basso LGM, Mendes LFS, Costa-Filho AJ. *Biophys Rev*. 2016; 8:179-191. Protein-membrane interactions play essential roles in a variety of cell functions such as signaling, membrane trafficking, and transport. Membrane-recruited cytosolic proteins that interact transiently and interfacially with lipid bilayers perform several of those functions. Experimental techniques capable of probing changes on the structural dynamics of this weak association are surprisingly limited. Among such techniques, electron spin resonance (ESR) has the enormous advantage of providing valuable local information from both membrane and protein perspectives by using intrinsic paramagnetic probes in metalloproteins or by attaching nitroxide spin labels to proteins and lipids. In this review, we discuss the power of ESR to unravel relevant structural and functional details of lipid-peripheral membrane protein interactions with special emphasis on local changes of specific regions of the protein and/or the lipids. First, we show how ESR can be used to investigate the direct interaction between a protein and a particular lipid, illustrating the case of lipid binding into a hydrophobic pocket of chlorocatechol 1,2-dioxygenase, a non-heme iron enzyme responsible for catabolism of aromatic compounds that are industrially released in the environment. In the second case, we show the effects of GPI-anchored tissue-nonspecific alkaline phosphatase, a protein that plays a crucial role in skeletal mineralization, and on the ordering and dynamics of lipid acyl chains. Then, switching to the protein perspective, we analyze the interaction with model membranes of the brain fatty acid binding protein, the major actor in the reversible binding and transport of hydrophobic ligands such as long-chain, saturated, or unsaturated fatty acids. Finally, we conclude by discussing how both lipid and protein views can be associated to address a common question regarding the molecular mechanism by which dihydroorotate dehydrogenase, an essential enzyme for the de novo synthesis of pyrimidine nucleotides, and how it fishes out membrane-embedded quinones to perform its function.

ABOUT THE EDITOR

Angel Catalá
Instituto de Investigaciones Fisicoquímicasteóricas y Aplicadas (INIFTA)
Facultad de Ciencias Exactas
Universidad Nacional de La Plata
La Plata, Argentina
catala@inifta.unlp.edu.ar

INDEX

#

α

A

B

C

D

E

F

G

H

I

J

K

L

Q

R

S

T

V

W